Springer Tracts in Natural Philosophy

Ergebnisse der angewandten Mathematik

Volume 1

Edited by C. Truesdell

Co-Editors: L. Collatz · G. Fichera

M. Fixman · P. Germain · J. Keller · A. Seeger

Linearized Analysis of
One - Dimensional
Magnetohydrodynamic Flows

Roy M. Gundersen

Springer-Verlag
Berlin · Göttingen · Heidelberg · New York
1964

Dr. Roy M. Gundersen
Professor of Mathematics
University of Wisconsin-Milwaukee

ISBN-13: 978-3-642-46007-4   e-ISBN-13: 978-3-642-46005-0
DOI: 10.1007/978-3-642-46005-0

© by Springer-Verlag · Berlin · Göttingen · Heidelberg 1964

Softcover reprint of the hardcover 1st edition 1964

Library of Congress Catalog Card Number 64-21654

Titel-Nr. 6729

Dedicated to my Parents

Preface

Magnetohydrodynamics is concerned with the motion of electrically conducting fluids in the presence of electric or magnetic fields. Unfortunately, the subject has a rather poorly developed experimental basis and because of the difficulties inherent in carrying out controlled laboratory experiments, the theoretical developments, in large measure, have been concerned with finding solutions to rather idealized problems. This lack of experimental basis need not become, however, a multimegohm impedance in the line of progress in the development of a satisfactory scientific theory. While it is true that ultimately a scientific theory must agree with and, in actuality, predict physical phenomena with a reasonable degree of accuracy, such a theory must be sanctioned by its mathematical validity and consistency. Physical phenomena may be expressed precisely and quite comprehensively through the use of differential equations, and the equations formulated by LUNDQUIST and discussed by FRIEDRICHS belong to a class of equations particularly well-understood and extensively studied. This class includes, in fact, many other eminent members, the solutions of which have led to results of far-reaching scientific and technological application. Frequently, the mathematical analysis has provided the foundations and guidance necessary for further developments, and, reciprocally, the physical problems have provided, in many cases, the impetus for the development of new mathematical theories which often have evolved to an a priori unpredictable extent.

In the fall of 1961, the author became seriously interested in the subject of magnetohydrodynamics, and this interest increased exponentially after reading the remarkable paper by FRIEDRICHS. Therein, FRIEDRICHS showed the amazingly close parallelism between the theory governing the motion of a compressible fluid whose electrical conductivity may be assumed to be infinite and the theory of conventional gas dynamics. Specifically, the hydromagnetic equations belong to the class of symmetric hyperbolic equations. Because these equations are also reducible, it is possible to develop a theory of shock waves and simple waves in a manner quite similar to that employed in ordinary gas dynamics. In this monograph, the author has used FRIEDRICHS' results as a basis for a more detailed examination of the mathematical structure of the equations governing a quite specialized class of hydromagnetic flows; namely, those which may be considered effectively one-dimensional or nearly so, subjected to a transverse magnetic field. Throughout, the emphasis is on the mathematical structure of the governing equations and their linearized form, and applications are made only to some well-known problems of conventional gas dynamics. The results of a little more than a year's research are presented and in no sense is completion even remotely claimed. If this work proves useful to workers in the field or stimulates others to direct their energies to this fascinating though as yet embryonic subject, the author will feel amply rewarded. Advancing technology requires that the period of gestation be reduced and waits for the birth and maturation of a more satisfactory theory.

A portion of this work was carried out while the author was in residence at the Mathematics Research Center, United States Army, University of Wisconsin. The author is indebted to Professor R. E. LANGER for making that stay possible and to the United States Army for support. Many of the results published herein first appeared in various articles published in the Journal of the Aerospace Sciences. The results of section 4.2 are reprinted with the permission of the Editors of the International Journal of Engineering Science.

The author wishes to thank Mrs. V. FIATTE for very expert typing and handwork on the equations.

Finally, the author wishes to thank Professor C. TRUESDELL for suggesting this monograph as a medium for presentation of the results and for facilitating its quite rapid publication.

Milwaukee, Wesconsin   ROY M. GUNDERSEN

September 1963

Table of Contents

Chapter 1

General Theory

1.1 Introduction

In common with the equations of conventional gas dynamics, the equations governing the motion of a perfectly conducting compressible fluid belong to the class of reducible hyperbolic equations. Because of this structural similarity, many of the techniques which have been applied so successfully to gas dynamic flows may be extended to magneto-hydrodynamic flows, and, in particular, it is possible to build up a theory of shock waves and simple waves remarkably parallel to that of conventional gas dynamics. This was pointed out quite elegantly by FRIEDRICHS [1] and amplified by others.

The equations of magnetohydrodynamics form a system of first order nonlinear partial differential equations. Because these equations can be solved exactly only in certain quite specialized cases, alternative approaches are mandatory. Whenever problems involving non-uniform shock waves are under consideration, entropy variations are involved and the differential equations for non-isentropic flow must be employed. Although non-isentropic flows are seldom readily amenable to exact mathematical analysis and the added effects of electrical conductivity only compound the difficulties, various perturbation techniques have led to important results and a better understanding of non-isentropic conventional gas dynamic flows. Therein lies the reason for this investigation.

The primary purpose of this monograph is the presentation of a method for discussing weakly non-isentropic quasi-one-dimensional flows of an ideal, inviscid, perfectly conducting compressible fluid subjected to a transverse magnetic field. It is the simplicity of application and the ease of extension which make the method attractive and, in fact, many of the solutions to the corresponding problems in conventional gas dynamic flows are contained as special cases, and this provides a valuable check on the theory. The analysis is based on characteristic perturbations developed by PAUL GERMAIN and the author for discussing similar problems for non-conducting fluids and the author's extensions of this work to magnetohydrodynamic flows.

First, the basic (exact) equations are written in characteristic form
and for isentropic flow, generalized Riemann invariants may be determined by quadrature. Further, a relation between the induction and the
local sound speed, valid for uniform or simple wave flows, is obtained
and used to reduce the governing equations from four to three. In order
to bring the equations into close analogy with those of non-conducting
fluids, a change of dependent variables is made so that the generalized
Riemann invariants are transformed into linear relations. When the
discussion is limited to weak entropy variations, the resultant system,
which in the limit of vanishing magnetic field reduces precisely to the
equations of conventional gas dynamics, may be linearized in the
neighborhood of a known (isentropic) state, viz., a uniform or simple
wave flow. A system of linear equations with the same characteristic
surfaces as the original system governs the terms of first order, and
solutions are obtained through the use of techniques first presented by
GERMAIN and GUNDERSEN [2].

Rather than unnecessarily lengthening the introduction, it seems
more efficient to discuss the problems and the pertinent literature in the
appropriate sections and to give only a brief outline of the problems
considered. As a first example of the use of the solution for the non-
isentropic perturbation of an initially uniform flow, the propagation of
non-uniform shock waves is considered. The non-uniform shocks may be
generated by considering the propagation of initially uniform shock
waves in ducts whose cross-sectional areas are made to undergo small
variations with both time and distance. While it is true that flows with
time-dependent area variations are rare (one example would be a
boundary layer growing with time), the time-dependence is included
since the corresponding solution follows immediately from that for only
spatial variations. Two basic examples are discussed; first, the effects of
non-uniform cross-sectional area variations on the propagation of an
initially uniform shock without regard for the method of generation
thereof save that whatever that may be, the mechanism is sufficiently
far removed so that no reflections come back to interfere with the basic
interaction of shock and area change. Then, the additional effects of
small non-uniformities in piston motion on a piston-driven shock propagating through a non-uniform duct are determined. In various con-
texts, these problems lead to differential relations between shock strength
and area change which, on integration, lead to area-shock strength
relationships valid for finite continuous area changes. By particular
choices of the area distributions, piston-driven converging cylindrical
and spherical shock waves may be discussed. The second example
concerns the effects due to small heat addition to an initially uniform
stream. Small entropy variations occur in this problem, and previous

work on the conventional gas dynamic case is extended to the magnetic case with little difficulty.

Simple wave flows are considered in some detail and, in particular, although simple waves do not exist in non-isentropic flow, it is shown that non-isentropic perturbations of simple waves in isentropic flow do exist. Further, from the simple wave theory, it is shown that the FRIEDRICHS theory of the formation and decay of shocks may be extended to the magnetic case.

Finally, some remarks are made on the effects of an oblique magnetic field.

1.2 The Linearized Equations

Since magnetohydrodynamics is concerned with the motion of electrically conducting fluids in the presence of electric or magnetic fields, the governing equations are determined by coupling MAXWELL's equations for the magnetic field with the usual equations of gas dynamics. The interaction process is exceedingly complicated since the motion of the fluid generates body forces which modify the motion and induced currents which modify the field. Throughout, it is assumed that the fluid is an ideal, polytropic, inviscid, non-heat conducting compressible continuum whose electrical conductivity is infinite. MAXWELL's displacements currents are neglected, and it is assumed that the medium is essentially neutral. Under these assumptions, the phenomenalistic laws may be assumed to be governed by the equations formulated by LUNDQUIST [3] and discussed by FRIEDRICHS [1].

It is a simple matter to modify these equations to take account of flows with area variations, and, in fact, the modifications appear only in the equation of continuity. Then, quasi-one-dimensional unsteady flow subjected to a transverse magnetic field is governed by the system of equations

$$P = \exp\left[(s - s^*)/c_v\right] \varrho^\gamma \tag{1.2.1}$$

$$\frac{2 c_t}{\gamma - 1} + \frac{2 u c_x}{\gamma - 1} + c u_x + \frac{c A_t}{A} + \frac{u c A_x}{A} = 0 \tag{1.2.2}$$

$$u_t + u u_x + \frac{2 c c_x}{\gamma - 1} + \frac{b^2 B_x}{B} - \frac{c^2 s_x}{\gamma (\gamma - 1) c_v} = 0 \tag{1.2.3}$$

$$B_t + u B_x + B u_x = 0 \tag{1.2.4}$$

$$s_t + u s_x = 0 \tag{1.2.5}$$

where partial derivatives are denoted by subscripts, and all dependent variables are functions of (x, t) alone. The definition of symbols is given in Appendix A.

1*

The system of Eqs. (1.2.2)—(1.2.5) is always hyperbolic, and the characteristics are given by

$$\frac{dx}{dt} = u,\; u + \omega,\; u - \omega$$

where ω, the limiting case of a fast wave, is the true speed of sound, i.e., small disturbances propagate with speed ω relative to the fluid. It is most convenient to deal with the characteristic form of the system where derivatives in only one direction appear in each equation. A complete derivation is given in Appendix B, and it is seen that for an arbitrary flow in a uniform duct, the system may be written in the following characteristic form

$$x_\beta - (u + \omega)\, t_\beta = 0 \tag{1.2.6}$$

$$x_\alpha - (u - \omega)\, t_\alpha = 0 \tag{1.2.7}$$

$$x_\xi - u t_\xi = 0 \tag{1.2.8}$$

$$\frac{(\omega^2 - c^2)\, B_\beta}{B} + \omega u_\beta + \frac{2 c c_\beta}{\gamma - 1} - \frac{c^2 s_\beta}{\gamma(\gamma - 1)\, c_v} = 0 \tag{1.2.9}$$

$$\frac{(\omega^2 - c^2)\, B_\alpha}{B} - \omega u_\alpha + \frac{2 c c_\alpha}{\gamma - 1} - \frac{c^2 s_\alpha}{\gamma(\gamma - 1)\, c_v} = 0 \tag{1.2.10}$$

$$\frac{B_\xi}{B} - \frac{2 c_\xi}{(\gamma - 1)\, c} = 0 \tag{1.2.11}$$

$$s_\xi = 0 . \tag{1.2.12}$$

Thus, three directions (parameterized by α, β, ξ) suffice to give four independent characteristic forms for the four basic equations. A rather surprising amount of information may be obtained from this characteristic system.

From Eq. (1.2.11), it follows that the quantity $B/c^{2/(\gamma-1)}$ is constant along each particle path, which is a well-known consequence of the assumption of infinite electrical conductivity, i.e., the magnetic field is "frozen" into the fluid, and for a constant state, this quantity is constant throughout the flow. Further, since a simple wave flow necessarily must be adjacent to a constant state from whence the particle paths originate, it is clear that $B/c^{2/(\gamma-1)}$ is also constant throughout a simple wave flow. Consequently, for an isentropic uniform or simple wave flow, i.e., the flows in the neighborhood of which perturbations will be considered, Eqs. (1.2.9) and (1.2.10) reduce to

$$u_\beta + \frac{2 \omega c_\beta}{(\gamma - 1)\, c} = 0 \tag{1.2.13}$$

$$- u_\alpha + \frac{2 \omega c_\alpha}{(\gamma - 1)\, c} = 0 . \tag{1.2.14}$$

Since $B = r_1 c^{2/(\gamma-1)}$, $\varrho = r_2 c^{2/(\gamma-1)}$, with constants r_1 and r_2,

$$b^2 = \frac{r_1^2 c^{2/(\gamma-1)}}{\mu r_2} \equiv k c^{2/(\gamma-1)}$$

so that

$$\omega^2 = c^2 \left[1 + k c^{\eta *}\right] . \qquad (1.2.15)$$

For a monatomic gas, $\gamma = 5/3$, and Eqs. (1.2.13) and (1.2.14) may be integrated explicitly to give

$$\frac{u}{2} + \frac{(1 + kc)^{3/2}}{k} = \frac{u}{2} + \frac{1}{k}\left(\frac{\omega}{c}\right)^3 = \alpha \qquad (1.2.16)$$

$$-\frac{u}{2} + \frac{(1 + kc)^{3/2}}{k} = -\frac{u}{2} + \frac{1}{k}\left(\frac{\omega}{c}\right)^3 = \beta \qquad (1.2.17)$$

while for an arbitrary value of γ, the solutions of Eqs. (1.2.13) and (1.2.14) may be written as

$$\frac{u}{2} + \frac{1}{(\gamma - 1)} \int (1 + kc^{\eta *})^{1/2}\, dc = \alpha \qquad (1.2.18)$$

$$-\frac{u}{2} + \frac{1}{(\gamma - 1)} \int (1 + kc^{\eta *})^{1/2}\, dc = \beta \qquad (1.2.19)$$

where (α, β) may be considered as generalizations of the usual Riemann invariants. The integral appearing in Eq. (1.2.18) may be expressed in terms of hypergeometric functions, but for most of the problems to be considered, an explicit evaluation is unnecessary. In the limit of vanishing magnetic field, Eqs. (1.2.18) and (1.2.19) may be integrated explicitly to yield the usual Riemann invariants, and these are linear relations between u and c.

In order to bring the governing equations (1.2.1)–(1.2.5) into close analogy with the equations of conventional gas dynamics, a change of dependent variables is made which transforms the generalized Riemann invariants into linear relations. This is effected by the change of variable $w = \int (1 + kc^{\eta *})^{1/2} dc$ which transforms Eqs. (1.2.2) and (1.2.3) to

$$\frac{w_t}{\gamma - 1} + \frac{u w_x}{\gamma - 1} + \frac{\omega u_x}{2} + \frac{u \omega A_x}{2A} + \frac{\omega A_t}{2A} = 0 \qquad (1.2.20)$$

$$\frac{u_t}{2} + \frac{u u_x}{2} + \frac{\omega w_x}{\gamma - 1} - \frac{c^2 s_x}{2\gamma(\gamma - 1) c_v} = 0 \qquad (1.2.21)$$

where use has been made of the relation $\dfrac{B_x}{B} = \dfrac{2 c_x}{(\gamma - 1) c}$ and Eq. (1.2.4) omitted. Adding and subtracting Eqs. (1.2.20) and (1.2.21) gives the system to be utilized in the sequel

$$\frac{w_t}{\gamma-1} + \frac{u_t}{2} + (u+\omega)\left[\frac{u_x}{2} + \frac{w_x}{\gamma-1}\right]$$

$$= -\frac{u\omega A_x}{2A} - \frac{\omega A_t}{2A} + \frac{c^2 s_x}{2\gamma(\gamma-1)c_v} \qquad (1.2.22)$$

$$\frac{w_t}{\gamma-1} - \frac{u_t}{2} + (u-\omega)\left[-\frac{u_x}{2} + \frac{w_x}{\gamma-1}\right]$$

$$= -\frac{u\omega A_x}{2A} - \frac{\omega A_t}{2A} - \frac{c^2 s_x}{2\gamma(\gamma-1)c_v} \qquad (1.2.23)$$

$$s_t + u s_x = 0 . \qquad (1.2.24)$$

The close analogy of this system with the basic equations of conventional gas dynamics should be noted, viz., a hyperbolic system of three non-linear first order partial differential equations with Riemann invariants which are linear relations between the dependent variables. This system includes the gas dynamic equations as the special case of vanishing magnetic field, i.e., for $k = 0$, $\omega = c$, $w = c$.

A formal linearization of Eqs. (1.2.22)—(1.2.24) in the neighborhood of a known (isentropic) state, i.e., a uniform or simple wave flow, denoted by the subscript zero, leads to the following system of linear equations, which has the same characteristic surfaces as Eqs. (1.2.22)—(1.2.24), for the terms of first order, denoted by the subscript one

$$R_t + (u_0 + \omega_0) R_x + (u_1 + \omega_1) \alpha_x$$

$$= -\frac{u_0\omega_0 A_{1x}}{2A_0} - \frac{\omega_0 A_{1t}}{2A_0} + \frac{c_0^2 s_{1x}}{2\gamma(\gamma-1)c_v} \qquad (1.2.25)$$

$$S_t + (u_0 - \omega_0) S_x + (u_1 - \omega_1) \beta_x$$

$$= -\frac{u_0\omega_0 A_{1x}}{2A_0} - \frac{\omega_0 A_{1t}}{2A_0} - \frac{c_0^2 s_{1x}}{2\gamma(\gamma-1)c_v} \qquad (1.2.26)$$

$$s_{1t} + u_0 s_{1x} = 0 \qquad (1.2.27)$$

where it has been noted that

$$\frac{u_0}{2} + \frac{w_0}{\gamma-1} = \alpha , \quad -\frac{u_0}{2} + \frac{w_0}{\gamma-1} = \beta$$

are the characteristic parameters of the base flow.

In the linearized form of the basic equations, the entropy equation may be solved directly and independently. From Eq. (1.2.27), it is clear that s_1 remains constant along the particle paths of the base flow, i.e., along $dx/dt = u_0$. Further, since $\varrho_0(dx - u_0 dt)$ is the exact differential of a function Ψ_0 such that Ψ_0 equated to a constant defines the trajectories, the solution of Eq. (1.2.27) may be written as

$$s_1 = \Gamma(\Psi_0)$$

with Γ an arbitrary differentiable function. It is convenient to define a function $T_0(x, t)$ by

$$c_0^2 s_{1x} = c_0^2 \varrho_0 \Gamma'(\Psi_0) = \gamma(\gamma - 1)c_v T_0 .$$

Since it follows from Eq. (1.2.15) that

$$\omega_1 = c_1 \left[\frac{\omega_0}{c_0} + \frac{\eta^*(\omega_0^2 - c_0^2)}{2\omega_0 c_0} \right]$$

the first order generalized Riemann invariants may be expressed in terms of u_1 and ω_1, viz.,

$$R = \frac{u_1}{2} + \frac{\omega_0^2 \omega_1}{[\omega_0^2 + (\gamma - 2)c_0^2]}$$

$$S = -\frac{u_1}{2} + \frac{\omega_0^2 \omega_1}{[\omega_0^2 + (\gamma - 2)c_0^2]} .$$

Thus

$$u_1 + \omega_1 = \left[\frac{3\omega_0^2 + (\gamma - 2)c_0^2}{2\omega_0^2} \right] R + \left[\frac{(\gamma - 2)c_0^2 - \omega_0^2}{2\omega_0^2} \right] S$$

$$u_1 - \omega_1 = \left[\frac{\omega_0^2 + (2 - \gamma)c_0^2}{2\omega_0^2} \right] R - \left[\frac{3\omega_0^2 + (\gamma - 2)c_0^2}{2\omega_0^2} \right] S .$$

Consequently, the non-isentropic perturbation of an initially uniform or simple wave flow may be determined by finding solutions of the following two first order linear equations for the first order generalized Riemann invariants

$$R_t + (u_0 + \omega_0)R_x + \left\{ \left[\frac{3\omega_0^2 + (\gamma - 2)c_0^2}{2\omega_0^2} \right] R + \left[\frac{(\gamma - 2)c_0^2 - \omega_0^2}{2\omega_0^2} \right] S \right\} \alpha_x$$

$$= -\frac{u_0 \omega_0 A_{1x}}{2A_0} - \frac{\omega_0 A_{1t}}{2A_0} + \frac{T_0}{2} \qquad (1.2.28)$$

$$S_t + (u_0 - \omega_0)S_x + \left\{ \left[\frac{\omega_0^2 + (2 - \gamma)c_0^2}{2\omega_0^2} \right] R - \left[\frac{3\omega_0^2 + (\gamma - 2)c_0^2}{2\omega_0^2} \right] S \right\} \beta_x$$

$$= -\frac{u_0 \omega_0 A_{1x}}{2A_0} - \frac{\omega_0 A_{1t}}{2A_0} - \frac{T_0}{2} . \qquad (1.2.29)$$

Although Eqs. (1.2.28) and (1.2.29) form a system of two coupled equations for the quantities R and S, these equations uncouple in each of the cases to be considered and may be solved separately. Since the equations are linear, solutions may be superposed. Further, the solutions may be obtained by solving the linear homogeneous system associated with Eqs. (1.2.28) — (1.2.29) and then adding a particular solution to the complete system in order to include the effects of cross-sectional and/or entropy perturbations.

For an initially uniform flow, α and β are constants, and the general solution of Eqs. $(1.2.28) - (1.2.29)$ is

$$R = F\left[x - (u_0 + \omega_0)\,t\right] +$$

$$+ \frac{E\left[x - u_0 t\right]}{\omega_0} - \frac{u_0 \omega_0 A_1\left[x, (u_0 + \omega_0)\, t/u_0\right]}{2 A_0\,(u_0 + \omega_0)} \qquad (1.2.30)$$

$$S = G\left[x - (u_0 - \omega_0)\,t\right] +$$

$$+ \frac{E\left[x - u_0 t\right]}{\omega_0} - \frac{u_0 \omega_0 A_1\left[x, (u_0 - \omega_0)\, t/u_0\right]}{2 A_0\,(u_0 - \omega_0)} \qquad (1.2.31)$$

$$T_0 = 2 E'\left[x - u_0 t\right] \qquad (1.2.32)$$

in terms of three arbitrary functions of one argument. From the general solution, the four distinct contributions to the perturbation may be noted; namely, one due to the entropy perturbation, which travels along the particle paths and is measured by E; a disturbance due directly to the area variations; a perturbation propagating with speed ω_0, the true speed of sound, with respect to the fluid along the family of characteristics $x - (u_0 + \omega_0)\,t = $ constant and measured by F; and a perturbation propagating with speed ω_0 with respect to the fluid, along the family of characteristics $x - (u_0 - \omega_0)\,t = $ constant and measured by G.

For the case of time-independent area variations, the general solution Eqs. $(1.2.30)-(1.2.32)$ reduces to that previously presented [4] for the corresponding problem. Since $u_1 = R - S$, it follows that the addition of an entropy perturbation affects w_1, i.e., c_1 and ω_1, but not u_1. Said differently, there exists a *particular* solution $u_1 = 0$ to the complete system. A general discussion of this phenomenon for conventional gas dynamic flows, including necessary and sufficient conditions for its occurrence, was given in [5] (see also [6]).

In the limit of vanishing magnetic field, the general solution Eqs. $(1.2.30)-(1.2.32)$ reduces exactly to the solution to the corresponding problem in conventional gas dynamics. Thus, there is the following result:

In the limit of vanishing magnetic field, the non-isentropic perturbation of an initially uniform one-dimensional flow of a perfectly conducting ideal compressible fluid, subjected to a transverse magnetic field, reduces to the solution for the corresponding problem for a non-conducting fluid.

Frequent physical and mathematical authentication of this result will be presented in the sequel. Although the proof is virtually trivial, the result is most definitely important. In particular, it shows that when the solution of the perturbation of an initially uniform hydromagnetic flow is determined, the solution of the corresponding problem in conventional gas dynamics is automatically contained as a special case.

Chapter 2

Shock Propagation in Non-Uniform Ducts
2.1 Introductory Comments

The most simple problem in shock dynamics is to determine the propagation of a uniform shock wave, i.e., a shock which separates two regions wherein the flow parameters are constant. Such a shock propagates with constant speed so that its path is represented in an (x, t) diagram by a straight line, and, since the entropy change through the shock is constant, only the differential equations for isentropic flow and the mechanical shock conditions need be considered in order to render the problem determinate. In many applications, however, it is necessary to solve problems involving non-uniform shock waves through which the entropy change, in general, is not constant. Since the speed of propagation of such a shock is not constant, its path is represented in an (x, t) diagram by a curve. Because of the entropy changes, the differential equations for non-isentropic flow would have to be solved, and this presents formidable difficulties. Consequently, the development of approximate analytical techniques which will yield the main features of the solution is a highly desirable goal. If the deviations from uniformity are not too large, i.e., if the resultant flows are only weakly non-isentropic, the present theory enables an approximate description to be given rather simply. In order to produce the non-uniform shocks, it is convenient to consider the propagation of initially uniform shock waves in ducts whose cross-sectional areas are made to undergo small variations. Since many of the results for conventional gas dynamic shock waves are contained as special cases of the present work and because of some continuing errors in the contemporary literature, it is of value to give a brief discussion of some of the previous work before explaining the method in detail.

The first results were obtained by CHESTER [8], who used a linearization based on small area variations and found that when an initially uniform plane shock wave of arbitrary strength passed through a non-uniform transition section, which joined two ducts of constant but unequal cross-sectional area, the concomitant pressure perturbation was given by

$$-K(P_2 - P_1)\,[\varDelta A]/A$$

where $P_2 - P_1$ was the initial pressure discontinuity across the shock, $[\varDelta A]$ the *net* change in cross-sectional area and the parameter K decreased monotonically with increasing shock strength. Although CHESTER started with the complete three-dimensional equations of motion, the exigencies of analysis made him utilize an averaging process so that the final results are restricted to the average pressure. But the average of a quantity over the cross-sectional area is what the hydraulic theory pre-dicts! Thus, it must be possible to bypass CHESTER's detailed analysis and the same results must be obtainable by investigating the problem from the very beginning by a one-dimensional formulation. This extremely important simplification, which provides the basis for many extensions of the theory, was observed by PAUL GERMAIN and the applicability of a quasi-one-dimensional analysis for discussing flows in non-uniform ducts noted in a preliminary presentation given by GERMAIN and GUNDER-SEN [2]. A more detailed presentation, including a derivation of CHESTER's results by the use of a purely one-dimensional analysis, was given by GUNDERSEN [5], [6].

A quite unexpected premium was the somewhat surprising result that the more simple one-dimensional theory led to an important improvement; namely, the term $[\varDelta A]$ in CHESTER's work was replaced by $\bar{A}$, the per-turbed area distribution. Consequently, instead of the pressure pertur-bation being related to the total change in cross-sectional area, a direct first order differential relation between changes in area and shock strength was valid at any point in the non-uniform section. The usefulness of this differential relation was exploited by CHISNELL [9], who was led to it differently under the assumption that the asymptotic form of CHESTER's steady state solution could be employed. CHISNELL integrated the dif-ferential relation in closed form and used the resultant shock strength-area relationship to give an approximate description of the motion of the shock in terms of the area of the duct. By suitable choices of the area distribution, a description of converging cylindrical and spherical shocks was given, and the results checked by comparison with previous simi-larity solutions, valid in the neighborhood of the points of collapse of the shocks. The comparison showed the remarkable accuracy of CHISNELL's work.

Later, WHITHAM [10] discussed these problems and their extensions and, in particular, rederived CHESTER's results by the use of a one-dimen-sional approach, virtually identical (the only difference being the choice of dependent variables) to that previously presented by GERMAIN and GUNDERSEN [2] and GUNDERSEN [5], [6]. ROSCISZEWSKI [11] also re-derived CHESTER's results, but made the erroneous statement in his paper that the one-dimensional approach to the problem was developed by WHITHAM. ROSCISZEWSKI limited credit to the present author's contri-bution to: "the influence of a slow perturbation of the otherwise uniform piston motion of a shock wave" though it seems somewhat curious that reference was made to the wrong paper, i.e., [12] instead of [6].

The aforementioned theories suffered from the common defect that they ceased to be valid for near-sonic flow behind the incident shock. By the use of techniques developed by STOCKER [13], this defect was removed in a discussion given by FRIEDMAN [14]. FRIEDMAN also incorrectly credited the one-dimensional approach to WHITHAM [10] though use was made of exactly the same dependent variables as in [2], [5] and [6]. Since FRIEDMAN's discussion is contained as a special case of a paper of GUNDERSEN [15] on secondary shock waves in magnetohydrodynamic channel flow, and this latter is merely a special case of the discussion of section 2.5, further comments may be deferred. It might be noted that [6], [10], [11], [12] and [14] appeared in the same journal.

The Germain-Gundersen theory was extended to discuss the propagation of a non-uniform shock wave into a moving fluid [16] and shock propagation in ducts with time and spatial area variations [7].

In general, the aforementioned papers dealt with the motion of an initially uniform shock wave without consideration of the method of generation, e.g., the uniform motion of a piston, save that the mechanism was sufficiently far removed so that no reflections came back to interfere with the basic interaction of the shock and the area change. Naturally, it was of interest to consider the additional effects of waves reflected back from a piston, and the perturbation of an initially uniform shock wave due to non-uniform piston motion in a uniform duct and in a duct with a linear area variation were given in [5], [6]. Extensions of CHISNELL's work, including closed-form shock strength-area relations, for piston-driven shocks were given for linear [12] and quadratic [17] area distributions, the appropriate ones for cylindrical and spherical shock waves. Piston-driven shock waves propagating through ducts with arbitrary area variations were considered by MIRELS [18] by a source distribution method and in [19] by the small perturbation method with the solution given as an infinite series. Further comments will be made in Chapter 3 which is devoted to piston-driven shocks.

The existence of explicit expressions for the generalized Riemann invariants for the one-dimensional unsteady flow of a perfectly conducting monatomic fluid subjected to a transverse magnetic field made it clear that it must be possible to extend the aforementioned work on the propagation of non-uniform shock waves to the magnetic case with little difficulty. It was shown [20] for the monatomic fluid that the perturbation generated when an initially uniform plane magnetohydrodynamic shock wave of arbitrary strength encountered an area variation could be determined, the problem being linearized on the basis of small area variations, and the solution presented in a form which included the conventional gas dynamic results, described in the foregoing, as a special case. The results of this analysis were used to develop a theory of converging cylindrical and spherical magnetohydrodynamic shock waves [21], and CHISNELL's work was contained as a special case. In [20], it was assumed that the fluid in front of the incident shock was at rest, but this work was extended to the case of non-stationary flow in front of the shock in [22]. Next, it was shown that the limitation of a monatomic fluid was unnecessary, in spite of the fact that generalized Riemann invariants could not be determined explicitly, at least in terms of elementary functions, and the theory of [20] and [21] was extended to an arbitrary value of the adiabatic index [4]. The extension of the results of [22] for arbitrary values of the adiabatic index is contained in the present work as a special case.

Before proceeding to the determination of the propagation of shocks in non-uniform ducts, it is of value to give a brief discussion of normal magnetohydrodynamic shock waves and the transition relations relating the flow parameters in front of and behind the shock. This is given in the next section.

2.2 Transition Relations Across Normal Shocks

Jump conditions across magnetohydrodynamic shock waves were considered by DE HOFFMAN and TELLER [23], LÜST [24], HELFER [25], FRIEDRICHS [1], BAZER and ERICSON [26], KANWAL [27], GUNDERSEN [28], [29] and others.

For a gas dynamic shock wave, the specification of the flow in front of and one quantity behind the shock suffices to determine the flow parameters behind the shock through the Rankine-Hugoniot conditions. For a hydromagnetic shock wave, two parameters are required; namely, one giving a measure of the shock strength and one giving a measure of the applied field. Then, all flow quantities may be expressed in terms of these two parameters and the known flow on one side of the shock wave.

Let the subscripts one and two denote, respectively, the flow in front of and behind the shock. Then, the hydromagnetic analogs of the Rankine-Hugoniot conditions are [1]

$$\varrho_1 v_1 = \varrho_2 v_2 , \tag{2.2.1}$$

$$B_1 v_1 = B_2 v_2 , \tag{2.2.2}$$

$$P_1 + \varrho_1 v_1^2 + \frac{B_1^2}{2\mu} = P_2 + \varrho_2 v_2^2 + \frac{B_2^2}{2\mu} \tag{2.2.3}$$

$$\frac{\gamma P_1}{(\gamma-1)\varrho_1} + \frac{v_1^2}{2} + \frac{B_1^2}{\varrho_1\mu} = \frac{\gamma P_2}{(\gamma-1)\varrho_2} + \frac{v_2^2}{2} + \frac{B_2^2}{\varrho_2\mu} \tag{2.2.4}$$

In terms of the ALFVÉN speed, relations (2.2.3) − (2.2.4) may be written as

$$P_1 + \frac{\varrho_1 b_1^2}{2} + \varrho_1 v_1^2 = P_2 + \frac{\varrho_2 b_2^2}{2} + \varrho_2 v_2^2 , \tag{2.2.5}$$

$$\frac{\gamma P_1}{(\gamma-1)\varrho_1} + \frac{v_1^2}{2} + b_1^2 = \frac{\gamma P_2}{(\gamma-1)\varrho_2} + \frac{v_2^2}{2} + b_2^2 . \tag{2.2.6}$$

The basic set to be discussed is Eqs. (2.2.1), (2.2.2), (2.2.5) and (2.2.6). It is assumed that the state one is completely known and further that the shock strength is prescribed. It is immediately apparent that

$$\frac{v_1}{v_2} = \frac{\varrho_2}{\varrho_1} = \frac{B_2}{B_1} = \frac{b_2^2}{b_1^2} = \sigma . \tag{2.2.7}$$

Eq. (2.2.7) may be used to express all quantities with subscript two in Eqs. (2.2.5) and (2.2.6) in terms of those with subscript one by the use of Eq. (2.2.7). This yields

$$\varrho_1 v_1^2 \left(1 - \frac{1}{\sigma}\right) + P_1 \left[1 - \tau + \frac{\gamma m_1^2 (1-\sigma)^2}{2}\right] = 0 , \tag{2.2.8}$$

$$\varrho_1 v_1^2 \left(1 - \frac{1}{\sigma^2}\right) + P_1 \left[\frac{2\gamma(1-\tau/\sigma)}{\gamma-1} + 2\gamma m_1^2(1-\sigma)\right] = 0 . \tag{2.2.9}$$

Since there exists a nontrivial solution to this linear system of homogeneous algebraic equations for P_1 and $\varrho_1 v_1^2$, the determinant of the coefficients must vanish, and this gives the result

$$m_1^2 = \frac{2\theta(\sigma - \tau) + 2(\sigma\tau - 1)}{\gamma(1-\sigma)^3}$$

which is linear in τ, so that

$$\tau = \frac{\theta\sigma - 1 - \gamma m_1^2(1-\sigma)^3/2}{\theta - \sigma} \tag{2.2.10}$$

which expresses τ in terms of σ and m_1.
Consequently,

$$\frac{c_2^2}{c_1^2} = \frac{\tau}{\sigma} = \frac{\theta\sigma - 1 - \gamma m_1^2(1-\sigma)^3/2}{\sigma(\theta - \sigma)} \tag{2.2.11}$$

$$\frac{m_2^2}{m_1^2} = \frac{\sigma^2}{\tau} = \frac{\sigma^2(\theta - \sigma)}{\theta\sigma - 1 - \gamma m_1^2(1-\sigma)^3/2} \tag{2.2.12}$$

$$\frac{\omega_2^2}{c_2^2} = 1 + m_2^2 = 1 + \frac{\sigma^2 m_1^2(\theta - \sigma)}{\theta\sigma - 1 - \gamma m_1^2(1-\sigma)^3/2} \tag{2.2.13}$$

$$\frac{\omega_2^2}{\omega_1^2} = \frac{(1 + m_2^2)\tau}{(1 + m_1^2)\sigma} \tag{2.2.14}$$

Expressing all quantities in Eqs. (2.2.5) and (2.2.6) in terms of those with subscript two gives

$$\varrho_2 v_2^2(\sigma - 1) + P_2 \left[\left(\frac{1}{\tau} - 1\right) + \frac{\gamma m_2^2}{2}\left(\frac{1}{\sigma^2} - 1\right)\right] = 0 \tag{2.2.15}$$

$$\varrho_2 v_2^2(\sigma^2 - 1) + P_2 \left[\frac{2\gamma}{(\gamma-1)}\left(\frac{\sigma}{\tau} - 1\right) + 2\gamma m_2^2\left(\frac{1}{\sigma} - 1\right)\right] = 0 . \tag{2.2.16}$$

The secular equation for the system of Eqs. (2.2.15) and (2.2.16) gives the known result $m_2^2/m_1^2 = \sigma^2/\tau$.

Division of Eq. (2.2.8) by $\varrho_1 c_1^2$ and Eq. (2.2.15) by $\varrho_2 c_2^2$ gives

$$M_1^2 = \frac{\sigma(1-\tau)}{\gamma(1-\sigma)} + \frac{m_1^2 \sigma(1+\sigma)}{2}$$

$$= \left\{ \frac{2}{\gamma-1} + m_1^2 \left[\frac{\gamma}{\gamma-1} + \frac{(2-\gamma)\sigma}{\gamma-1} \right] \right\} \frac{\sigma}{(\theta-\sigma)} \tag{2.2.17}$$

$$M_2^2 = \frac{M_1^2}{\sigma\tau}. \tag{2.2.18}$$

Further, there is the result

$$\frac{v_1^2}{\omega_1^2} = \frac{M_1^2}{1+m_1^2}$$

$$\frac{v_2^2}{\omega_2^2} = \frac{M_2^2}{1+m_2^2}.$$

From Eq. (2.2.1),

$$\frac{u_2}{c_1} = \frac{U-u_1+u_1}{c_1} = \frac{M_1}{\sigma}$$

so that

$$n_2 = \frac{u_2}{c_2} = \left[n_1 + \left(1 - \frac{1}{\sigma}\right) M_1 \right] \left(\frac{\sigma}{\tau} \right)^{\frac{1}{2}}$$

or

$$n_2 = \left[n_1 + M_1 \left(1 - \frac{1}{\sigma}\right) \right] \left[\frac{\sigma(\theta-\sigma)}{\theta\sigma - 1 - \gamma m_1^2 (1-\sigma)^3/2} \right]^{\frac{1}{2}}. \tag{2.2.19}$$

Finally

$$\frac{u_1}{\omega_1} = \frac{n_1}{(1+m_1^2)^{\frac{1}{2}}}$$

$$\frac{u_2}{\omega_2} = \frac{n_2}{(1+m_2^2)^{\frac{1}{2}}}.$$

Consequently, all quantities behind the shock have been expressed in terms of the quantities in front of the shock and the two parameters m_1, which is a measure of the applied field, and σ, the shock strength. The usual gas dynamic relations are obtained by letting $m_1 = 0$.

The existence of an applied magnetic field produces a magnetic pressure counteracting the gas pressure such that the sum of the gas pressure and the magnetic pressure is constant. As a consequence, there is a decrease in the rate of momentum flow of the fluid, and the magnetic field is "frozen" into and compressed to the same extent as the fluid and causes a decrease in compression across a shock wave. From Eq. (2.2.10) it is also immediately apparent that for very weak or very strong shock waves [in an asymptotic sense], the shock strength is independent of the applied field. The solutions obtained in the subsequent sections exhibit exactly this behavior.

Eq. (2.2.10) gives, in fact, the magnetic Hugoniot function for polytropic fluids, i.e.,

$$P_2(\varrho_2 - \theta\varrho_1) - \left[\varrho_1 - \theta\varrho_2 + \frac{\gamma m_1^2}{2}\varrho_1\left(1 - \frac{\varrho_2}{\varrho_1}\right)^3\right]P_1 = 0 . \qquad (2.2.20)$$

Along the Hugoniot curve, τ varies from 0 to ∞, while σ varies between the limits $\sigma_{\min} = \theta$ and $\sigma_{\max}$, where $\sigma_{\max}$ is determined from the equation

$$\theta\sigma - 1 - \frac{\gamma m_1^2}{2}(1 - \sigma)^3 = 0 ;$$

along the Hugoniot curve, $\dfrac{d\tau}{d\sigma} > 0$.

2.3 Solution behind the Incident Shock

The problem considered is that of an initially uniform plane magnetohydrodynamic shock wave of arbitrary strength propagating with initially constant speed in a duct which has a section of non-uniform cross-sectional area, and it is assumed that the area variations are spatially and time-dependent. It is assumed that the (unspecified) mechanism of generation of the shock is sufficiently far removed so that no reflected waves come back to interfere with the basic interaction to be considered. It is assumed that the area variations are restricted to the region $x > 0$, whereas to the left of the cross-section $x = 0$, the tube is of constant cross-sectional area, and the shock propagates therein with constant speed. Since the area variations are considered to be time-dependent, the fluid in front of the shock is perturbed, but, to simplify the discussion, it is assumed that no disturbances

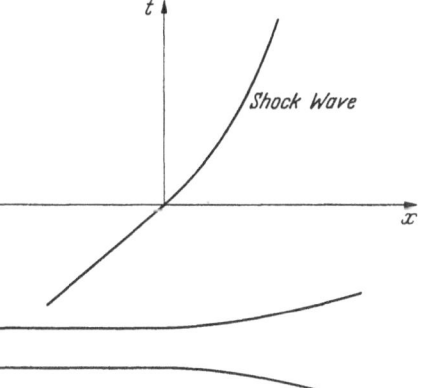

Fig. 1. The (x, t) diagram for the shock wave area interaction. To the left of the cross-section $x = 0$, the duct is of constant area, and the shock propagates with uniform speed in this portion of the duct. It is assumed that no perturbations from the disturbed upstream flow reach the shock until the section $x = 0$ is reached at $t = 0$

from the perturbed upstream flow reach the shock until the area variations are encountered at $x = 0$ at $t = 0$. The set-up of the problem is given in Fig. 1.

Initially, the flow behind the incident shock is a uniform, isentropic state, but, when the shock wave meets the area variation, the shock is perturbed, its strength altered and the subsequent flow non-isentropic. Although the one-dimensional approach does not allow a complete description of the complex metamorphosis which takes place behind the shock due to the reflections of rarefactive or compressive wavelets by the non-uniformities, the gross features of the interaction process are readily discernible. There are three distinct contributions to the disturbance; namely, a permanent disturbance due directly to the area change; a transient disturbance, which propagates with speed ω (the true sonic speed), with respect to the flow behind the shock, and is due to reflections of the permanent perturbation at the shock; and the effect of the perturbed flow in front of the shock.

In the immediate vicinity of the incident shock, the ultimate effect is an altered shock strength and concomitant pressure change behind the shock. When the main flow behind the shock has speed $< \omega$, a transient reflected disturbance is convected to the left downstream of the shock with speed ω relative to the fluid; for flow speed $> \omega$, this reflected disturbance is convected to the right upstream of the shock with speed ω relative to the main flow. Expressions for these various contributions are obtained.

Under the assumption of small area variations, the resultant flow behind the incident shock wave will be weakly non-isentropic and may be determined by use of the solution for the non-isentropic perturbation of a constant state, Eqs. (1.2.30)−(1.2.32). In the solution for R, the term $F[x − (u_2 + \omega_2)t]$ would represent a disturbance wave travelling to the right from downstream of the shock. But from the formulation of the problem, there is no mechanism downstream of the shock which could give rise to such a term; thus, the pressure perturbation immediately behind the incident shock may be obtained simply by setting $F = 0$, and this is carried out in the present section. Once this solution has been determined, it may be used to evaluate the arbitrary function G in the solution for S, and this determines the transient reflected disturbance; the solution is given in the next section.

Consequently, setting $F \equiv 0$ in Eq. (1.2.30) gives

$$\frac{\bar{u}_2}{2} + \frac{\bar{w}_2}{\gamma - 1} - \frac{c_2^2 \bar{s}_2}{2 \omega_2 \gamma (\gamma - 1) c_v} = \frac{-u_2 \omega_2 \bar{A}_2 [x, (n_2 + q_2) t/n_2]}{2 A_2 (u_2 + \omega_2)} \quad (2.3.1)$$

where perturbations of a base quantity are denoted by a bar. Since

$$\bar{w}_2 = q_2 \bar{c}_2 ;$$

$$\frac{-\bar{s}_2}{2 \gamma (\gamma - 1) c_v} = \frac{\bar{P}_2}{2 \gamma P_2} - \frac{\bar{c}_2}{(\gamma - 1) c_2} ,$$

the latter from Eq. (1.2.1), Eq. (2.3.1) may be written as

$$\frac{\bar{u}_2}{c_2} + \left[\frac{\gamma (m_2^2 + 1) - 1}{\gamma (\gamma - 1) q_2} \right] \frac{\bar{P}_2}{P_2} - \frac{m_2^2}{(\gamma - 1) q_2} \frac{\bar{\varrho}_2}{\varrho_2}$$
$$= \frac{-n_2 q_2 \bar{A}_2 [x, (n_2 + q_2) t/n_2]}{(n_2 + q_2) A_2} . \quad (2.3.2)$$

From the perturbation form of the jump conditions of section 2.2

$$\frac{\bar{\tau}}{\tau} = \frac{\bar{P}_2}{P_2} - \frac{\bar{P}_1}{P_1} =$$

$$\left\{ \frac{(\theta^2 - 1)\,\sigma + m_1^2\,\gamma\sigma\,[3\theta - 1 - 6\theta\sigma + 3\,(\theta + 1)\,\sigma^2 - 2\sigma^3]/2}{(\theta - \sigma)\,[\theta\sigma - 1 - \gamma m_1^2(1 - \sigma)^3/2]} \right\} \frac{\bar{\sigma}}{\sigma} \qquad (2.3.3)$$

$$- \frac{\gamma m_1\,(1 - \sigma)^3\,\bar{m}_1}{\theta\sigma - 1 - \gamma m_1^2(1 - \sigma)^3/2}$$

$$\frac{\bar{u}_2}{c_2} = n_1 \left(\frac{\sigma}{\tau}\right)^{\frac{1}{2}} \frac{\bar{u}_1}{u_1} + \left[n_2 - n_1 \left(\frac{\sigma}{\tau}\right)^{\frac{1}{2}} \right] \frac{\bar{c}_1}{c_1} +$$

$$+ \frac{\left[n_2 - n_1 \left(\frac{\sigma}{\tau}\right)^{\frac{1}{2}} \right] \left[1 + \theta + \sigma\,(\theta - 3) \right] m_1 \bar{m}_1}{2\,(\theta + 1)/\gamma + m_1^2\,[1 + \theta + \sigma\,(\theta - 3)]} + \left[n_2 - n_1 \left(\frac{\sigma}{\tau}\right)^{\frac{1}{2}} \right] \times \qquad (2.3.4)$$

$$\times \left\{ \frac{2\,(\theta + 1)\,\theta/\gamma + 2\,(\theta + 1)\,(\theta - 2)\,\sigma/\gamma + m_1^2[\theta\,(1 + \theta) + (1 + \theta)\,(\theta - 2)\,\sigma + (\theta - 3)\,(2\theta - 1)\,\sigma^2 + (3 - \theta)\,\sigma^3]}{2\,(\theta - \sigma)\,(\sigma - 1)\,\{2\,(\theta + 1)/\gamma + m_1^2\,[1 + \theta + \sigma\,(\theta - 3)]\}} \right\} \frac{\bar{\sigma}}{\sigma}\,.$$

Thus, substituting Eqs. (2.3.3)—(2.3.4) into Eq. (2.3.2) gives

$$\left\{ \left[n_2 - n_1 \left(\frac{\sigma}{\tau}\right)^{\frac{1}{2}} \right] \times \right.$$

$$\times \left[\frac{2\,(\theta + 1)\,\theta/\gamma + 2\,(\theta + 1)\,(\theta - 2)\,\sigma/\gamma + m_1^2[\theta\,(1 + \theta) + (1 + \theta)\,(\theta - 2)\,\sigma + (\theta - 3)\,(2\theta - 1)\,\sigma^2 + (3 - \theta)\,\sigma^3]}{2\,(\sigma - 1)\,\{2\,(\theta + 1)/\gamma + m_1^2\,[1 + \theta + \sigma\,(\theta - 3)]\}} \right] +$$

$$+ \left[\frac{\gamma\,(m_2^2 + 1) - 1}{\gamma\,(\gamma - 1)\,q_2} \right] \left[\frac{(\theta^2 - 1)\,\sigma + m_1^2\,\gamma\sigma\,[3\theta - 1 - 6\theta\sigma + 3\,(\theta + 1)\,\sigma^2 - 2\sigma^3]/2}{\theta\sigma - 1 - \gamma m_1^2(1 - \sigma)^3/2} \right] -$$

$$- \left. \frac{m_2^2\,(\theta - \sigma)}{q_2\,(\gamma - 1)} \right\} \frac{\bar{P}_2}{(P_2 - P_1)}$$

$$- - \left[\frac{(\theta^2 - 1)\,\sigma + m_1^2\,\gamma\sigma\{3\theta - 1 - 6\theta\sigma + 3\,(\theta + 1)\,\sigma^2 - 2\sigma^3\}/2}{(\theta + 1)\,(\sigma - 1) + \gamma m_1^2(\sigma - 1)^3/2} \right] \times$$

$$\times \left\{ \frac{n_2 q_2 \bar{A}_2}{(n_2 + q_2)\,A_2} + n_1 \left(\frac{\sigma}{\tau}\right)^{\frac{1}{2}} \frac{\bar{u}_1}{u_1} + \left[n_2 - n_1 \left(\frac{\sigma}{\tau}\right)^{\frac{1}{2}} \right] \frac{\bar{c}_1}{c_1} - \frac{m_2^2}{(\gamma - 1)\,q_2} \frac{\bar{\varrho}_1}{\varrho_1} \right\} -$$

$$- \left[n_2 - n_1 \left(\frac{\sigma}{\tau}\right)^{\frac{1}{2}} \right] \frac{\bar{m}_1}{m_1} \times$$

$$\times \left[\left\{ \frac{2\,(\theta + 1)\,\theta/\gamma + 2\,(\theta + 1)\,(\theta - 2)\,\sigma/\gamma + m_1^2[\theta\,(1 + \theta) + (1 + \theta)\,(\theta - 2)\,\sigma + (\theta - 3)\,(2\theta - 1)\,\sigma^2 + (3 - \theta)\,\sigma^3]}{2\,(\theta - \sigma)\,(\sigma - 1)\,\{2\,(\theta + 1)/\gamma + m_1^2\,[1 + \theta + \sigma\,(\theta - 3)]\}} \right. \right.$$

$$- \left. \frac{m_2^2}{q_2\,(\gamma - 1)} \right\} \frac{\gamma m_1^2\,(1 - \sigma)^3\,(4 - \sigma)}{(\theta + 1)\,(\sigma - 1) + \gamma m_1^2(\sigma - 1)^3/2} +$$

$$+ \frac{[1 + \theta + \sigma\,(\theta - 3)]\,m_1^2}{2\,(\theta + 1)/\gamma + m_1^2[1 + \theta + \sigma\,(\theta - 3)]} \times$$

$$\times \left. \left\{ \frac{(\theta^2 - 1)\,\sigma + m_1^2\,\gamma\sigma\,[3\theta - 1 - 6\theta\sigma + 3\,(\theta + 1)\,\sigma^2 - 2\sigma^3]/2}{(\theta + 1)\,(\sigma - 1) + \gamma m_1^2(\sigma - 1)^3/2} \right\} \right] +$$

$$+ \left[n_2 - n_1 \left(\frac{\sigma}{\tau}\right)^{\frac{1}{2}} \right] \frac{\bar{P}_1}{P_1}\,(4 - \sigma) \left[\frac{\gamma m_1^2\,(1 - \sigma)^3/2 + 1 - \theta\sigma}{(\theta + 1)\,(1 - \sigma) + \gamma m_1^2(1 - \sigma)^3/2} \right] \times$$

$$\times \left\{ \frac{2\,\theta\,(\theta + 1)/\gamma + 2\,(\theta + 1)\,(\theta - 2)\,\sigma/\gamma + m_1^2[\theta\,(1 + \theta) + (1 + \theta)\,(\theta - 2)\,\sigma + (\theta - 3)\,(2\theta - 1)\,\sigma^2 + (3 - \theta)\,\sigma^3]}{2\,(\theta - \sigma)\,(\sigma - 1)\,\{2\,(\theta + 1)/\gamma + m_1^2\,[1 + \theta + \sigma\,(\theta - 3)]\}} \right. -$$

$$- \left. \frac{m_2^2}{q_2\,(\gamma - 1)} \right\}$$

which may be abbreviated as:

$$\bar{P}_2/(P_2 - P_1) = - K(\sigma, m_1, n_1)\, \bar{A}_2\,[x, (n_2 + q_2)\,t/n_2]/A_2 +$$
$$+ \sum_{i=1}^{5} D_i(\sigma, m_1, n_1)\, \phi_i(x, t) \tag{2.3.5}$$

where $\phi_1 = \bar{\varrho}_1/\varrho_1$, $\phi_2 = \bar{u}_1/u_1$, $\phi_3 = \bar{c}_1/c_1$, $\phi_4 = \bar{P}_1/P_1$, $\phi_5 = \bar{m}_1/m_1$ and K and D_i refer to the respective coefficients. Eq. (2.3.5) gives the pressure perturbation immediately behind the incident shock. The term $-K\bar{A}_2/A_2$ gives the contribution due directly to the area variations while the summation term gives that due to the perturbed upstream flow.

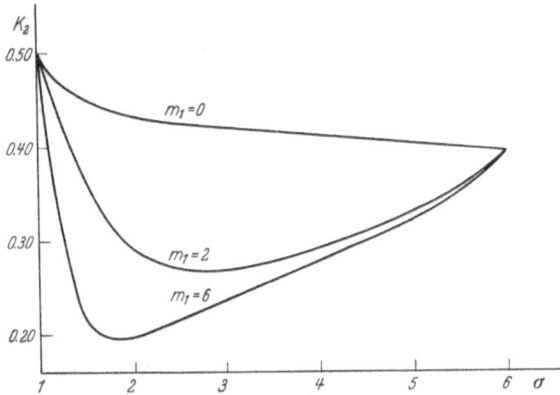

Fig. 2. A graph of the parameter $K(\sigma, m_1, 0) \equiv K_2$ vs. σ for $m_1 = 0$, the ordinary gas dynamic case, and for $m_1 = 2$ and 6

For time-independent ducts, i.e., $\bar{A}_2 = \bar{A}_2(x)$ alone, Eq. (2.3.5) contains as a special case the results given [22] for the propagation of a non-uniform hydromagnetic shock wave into a moving monatomic fluid; with $n_1 = 0$ and $\phi_i = 0, i = 1 - 5$, Eq. (2.3.5) reduces to the solution given [4] for the propagation of a non-uniform shock wave into a stationary fluid. In the limit of vanishing magnetic field, Eq. (2.3.5) reduces exactly to the corresponding solution for the propagation of a conventional gas dynamic shock wave in a duct with time-dependent area variations [7].

In magnetohydrodynamics, strong shocks can occur in two ways, viz., for σ close to θ, i.e., 6 for $\gamma = 7/5$, or for a very strong applied field for any $\sigma > 1$, i.e., m_1 is then large.

For stationary flow in front of the incident shock wave, it is seen that for $\gamma = 7/5$ and all m_1, $\lim_{\sigma \to 1+} K(\sigma, m_1, 0) = 0.5$ and $\lim_{\sigma \to 6-} K(\sigma, m_1, 0) = 0.39414$, and these are exactly the limits for the ordinary gas dynamic case which corresponds to $m_1 = 0$. Consequently, there is the result:

For very strong or very weak shock waves [in an asymptotic sense], the propagation of the shock is independent of the applied field and given by conventional gas dynamic theory.

For $m_1 = 0$, $K(\sigma, m_1, 0)$ decreases monotonically with increasing σ and agrees exactly with the corresponding parameter in the gas dynamic case. For any $m_1 \neq 0$, there is no longer a monotonic variation with σ,

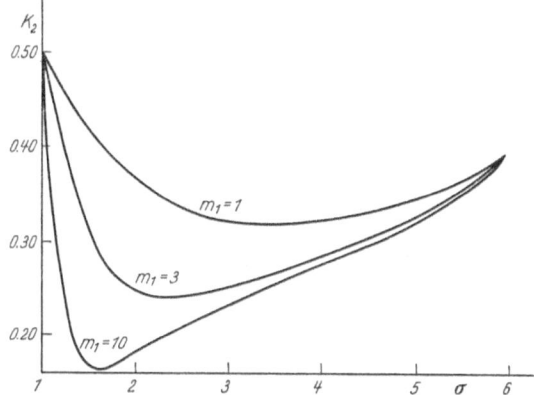

Fig. 3. A graph of the parameter K_2 vs. σ for $m_1 = 1, 3$ and 10

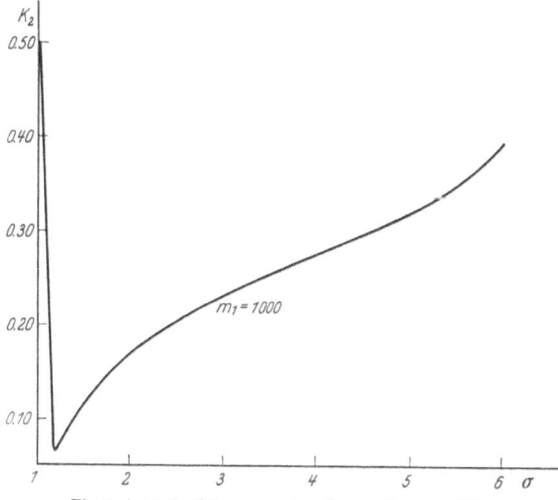

Fig. 4. A graph of the parameter K_2 vs. σ for $m_1 = 1000$

but the curves are concave upward with those for greater m_1 lying beneath those for lesser m_1, and all curves issue from the point $(\sigma, K) = (1, 0.5)$ and terminate at the point $(6, 0.39414)$. Graphs are presented in Figs. 2, 3 and 4.

For fixed incident shock strength σ, $K(\sigma, m_1, 0)$ decreases monotonically with increasing m_1, so that in a diverging (converging) duct, the pressure decrement (increment) is decreased (decreased) by increasing the applied field. Qualitatively, the motion of the shock is independent of whether the main flow behind the shock is $< \omega_2$ or $> \omega_2$.

For non-stationary flow in front of the incident shock, a complete discussion is quite difficult since the pressure perturbation immediately downstream of the shock is the sum of two contributions, viz., the direct effect of the area variations and the perturbed flow upstream. The latter contribution, which depends implicitly on the area variations, will not be discussed, but the effect of the behavior of the former on the flow may be obtained from the tables presented in Appendix C.

For fixed n_1, $K(\sigma, 0, n_1)$ decreases monotonically with increasing σ, i.e., the same essential behavior as $K(\sigma, 0, 0)$. Since $K(\sigma, 0, n_1)$ increases monotonically with increasing n_1 for fixed σ, in a diverging (converging) duct, this portion of $\overline{P}_2$, the pressure decrement (increment), which decreases (increases) monotonically with increasing n_1, is increased (increased) by increasing the Mach number of the upstream flow. For fixed n_1 and $m_1 \neq 0$, K no longer varies monotonically with increasing σ, but the curves are concave upward with curves for greater m_1 lying beneath those for lesser m_1. For large m_1, the minimum point of each curve lies between $\sigma = 1$ and $\sigma = 1.1$ so that the tables show only a monotonic increase with increasing σ. For fixed σ (and n_1), K decreases monotonically with increasing m_1, i.e., the same essential behavior as $K(\sigma, m_1, 0)$. For fixed m_1 and σ, $K(\sigma, m_1, n_1)$ increases monotonically with n_1.

2.4 The Reflected Disturbance

To complete the solution for the perturbed flow behind the incident shock wave, an expression for the transient reflected disturbance must be obtained from Eq. (1.2.31). The evaluation of the arbitrary function G is accomplished quite readily by noting that the system Eqs. (1.2.30) — (1.2.32) may be written as:

$$\frac{\bar{u}_2}{c_2} + \frac{2 m_2^2}{(\gamma - 1) q_2} \frac{\bar{c}_2}{c_2} + \frac{\overline{P}_2}{\gamma P_2 q_2} = - \frac{n_2 q_2}{(n_2 + q_2) A_2} \overline{A}_2 \left[x, \frac{(n_2 + q_2) t}{n_2} \right]$$

$$- \frac{\bar{u}_2}{2} + \frac{2 m_2^2}{(\gamma - 1) q_2} \frac{\bar{c}_2}{c_2} + \frac{\overline{P}_2}{\gamma P_2 q_2} = - \frac{n_2 q_2}{(n_2 - q_2) A_2} \overline{A}_2 \left[x, \frac{(n_2 - q_2) t}{n_2} \right] +$$

$$+ \frac{2 G}{c_2} \left[x - (u_2 - \omega_2) t \right] .$$

Adding these two equations, which serves to eliminate $\bar{u}_2$, using Eq. (2.3.3) and the fact

$$\frac{\bar{c}_2}{c_2} - \frac{\bar{c}_1}{c_1} = \frac{\bar{\tau}}{2\tau} - \frac{\bar{\sigma}}{2\sigma} ,$$

it follows that:

$$\frac{\overline{P}_2}{P_2}\left[\frac{\gamma-1+\gamma m_2^2}{\gamma(\gamma-1)q_2}-\right.$$

$$-\frac{m_2^2(\theta-\sigma)\,[\theta\sigma-1-\gamma m_1^2(1-\sigma)^3/2]}{(\gamma-1)\,q_2[(\theta^2-1)\,\sigma+m_1^2\gamma\sigma\{3\theta-1-6\theta\sigma+3(\theta+1)\sigma^2-2\sigma^3\}/2]}\Bigg]-$$

$$-\frac{\overline{P}_1}{P_1}\,\frac{m_2^2}{(\gamma-1)\,q_2}\times$$

$$\times\left[1-\frac{(\theta-\sigma)\,[\theta\sigma-1-\gamma m_1^2(1-\sigma)^3/2]}{(\theta^2-1)\,\sigma+m_1^2\,\gamma\sigma\,[3\theta-1-6\theta\sigma+3(\theta+1)\,\sigma^2-2\sigma^3]/2}\right]-$$

$$-\frac{m_2^2\,\gamma m_1\,(1-\sigma)^3\,(\theta-\sigma)\,\overline{m}_1}{(\gamma-1)\,q_2\{(\theta^2-1)\,\sigma+m_1^2\,\gamma\sigma\,[3\theta-1-6\theta\sigma+3(\theta+1)\,\sigma^2-2\sigma^3]/2\}}+$$

$$+\frac{2\,m_2^2}{(\gamma-1)\,q_2}\,\frac{\overline{c}_1}{c_1}+\frac{n_2q_2}{(n_2+q_2)}\,\frac{\overline{A}_2\,[x,\,(n_2+q)\,t/n_2]}{2A_2}+$$

$$+\frac{n_2q_2}{n_2-q_2}\,\frac{\overline{A}_2\,[x,\,(n_2-q_2)\,t/n_2]}{2A_2}=G\,[x-(u_2-\omega_2)\,t]/c_2\,.$$

Evaluating this on the shock, $x=Ut$, and replacing $\overline{P}_2/P_2$ by its value from Eq. (2.3.5) evaluates G so that the final solution is:

$$\overline{P}_2/(P_2-P_1)=-2m_2^2\nu_1\,[\phi_3(x,\,t)-\phi_3(\delta\lambda,\,\zeta\lambda)]/\nu_2+$$

$$+\sum_{i=1}^{5}D_i(\sigma,m_1,\,n_1)\,\phi_i\,[\delta\lambda,\,\zeta\lambda]+$$

$$+m_2^2\left[1-\frac{(\theta-\sigma)\,[\theta\sigma-1-\gamma m_1^2(1-\sigma)^3/2]}{(\theta^2-1)\,\sigma+m_1^2\,\gamma\sigma\,[3\theta-1-6\theta\sigma+3(\theta+1)\,\sigma^2-2\sigma^3]/2}\right]\times$$

$$\times\frac{\nu_1}{\nu_2}\cdot[\phi_4(x,\,t)-\phi_4(\delta\lambda,\zeta\lambda)]+$$

$$+\left[\frac{\gamma m_2^2m_1^2(1-\sigma)^3\,(\theta-\sigma)}{(\theta^2-1)\,\sigma+m_1^2\,\gamma\,\sigma\,[3\theta-1-6\theta\sigma+3(\theta+1)\,\sigma^2-2\sigma^3]/2}\right]\times\qquad(2.4.1)$$

$$\times\frac{\nu_1}{\nu_2}\,[\phi_5(x,\,t)-\phi_5(\delta\lambda,\zeta\lambda)]+$$

$$+\frac{q_2^2n_2\,(\gamma-1)\,\nu_1}{2\nu_2(n_2+q_2)\,A_2}\,\{\overline{A}_2[\delta\lambda,\,(n_2+q_2)\,\zeta\lambda/n_2]-\overline{A}_2\,[x,\,(n_2+q_2)\,t/n_2]\}+$$

$$+\frac{q_2^2n_2(\gamma-1)\,\nu_1}{2\nu_2(n_2-q_2)\,A_2}\,\{\overline{A}_2[\delta\lambda,\,(n_2-q_2)\,\zeta\lambda/n_2]-\overline{A}_2\,[x,\,(n_2-q_2)\,t/n_2]\}-$$

$$-K(\sigma,\,m_1,\,n_1)\,\overline{A}_2[\delta\lambda,\,(n_2+q_2)\,\zeta\lambda/n_2]/A_2$$

where

$$\nu_1 = \frac{\theta\sigma - 1 - \gamma m_1^2 (1 - \sigma)^3/2}{(\theta + 1)(\sigma - 1) + \gamma m_1^2 (\sigma - 1)^3/2}$$

$$\nu_2 = \frac{\gamma - 1 + \gamma m_2^2}{\gamma} - \frac{m_2^2 (\theta - \sigma)[\theta\sigma - 1 - \gamma m_1^2 (1 - \sigma)^3/2]}{(\theta^2 - 1)\sigma + m_1^2 \gamma\sigma[3\theta - 1 - 6\theta\sigma + 3(\theta + 1)\sigma^2 - 2\sigma^3]/2}$$

$$\lambda = x - (u_2 - \omega_2)t$$

$$\delta = \frac{M_2 + n_2}{M_2 + q_2}$$

$$\zeta = \frac{1}{c_2[M_2 + q_2]}$$

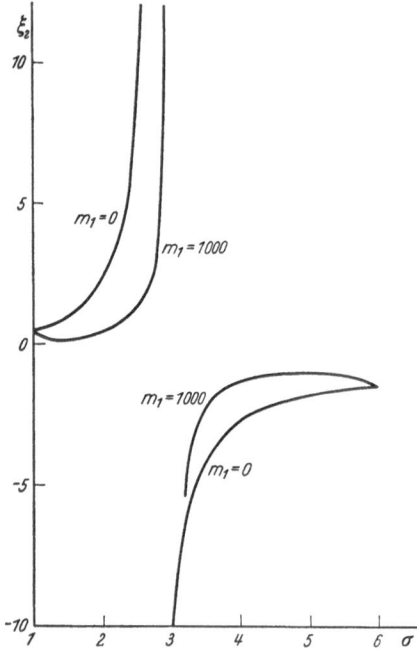

Fig. 5. A graph of the parameter $Z(\sigma, m_1, 0) \equiv \xi_2$ vs. σ. The parameter is singular for true-sonic flow behind the incident shock, i.e., for $n_2 = q_2$, and this singularity masks the formation of secondary shocks

When the flow speed behind the shock is $< \omega_2$, a disturbance will be reflected to the left downstream of the shock, and the pressure perturbation will ultimately be given by Eq. (2.4.1) with the two terms $\bar{A}_2[x, (n_2 \pm q_2)t/n_2]$ absent since $\bar{A}_2 = 0$ for $x < 0$. For flow speed $> \omega_2$ behind the incident shock, the reflected disturbance will be convected to the right upstream of the shock and is given by Eq. (2.4.1). When the area variations are assumed time-independent, Eq. (2.4.1) reduces to

$$\bar{P}_2/(P_2 - P_1) = \Phi_4(x, t, \lambda) - \frac{(\gamma - 1) n_2^2 q_2^2 \nu_1 \bar{A}_2(x)}{(n_2^2 - q_2^2)\nu_2 A_2} -$$

$$- \left\{ K(\sigma, m_1, n_1) - \frac{(\gamma - 1) n_2^2 q_2^2 \nu_1}{(n_2^2 - q_2^2)\nu_2} \right\} \frac{\bar{A}_2(\delta\lambda)}{A_2} \equiv \qquad (2.4.2)$$

$$\equiv \Phi_4(x, t, \lambda) - Y(\sigma, m_1, n_1)\bar{A}_2(x)/A_2 - Z(\sigma, m_1, n_1)\bar{A}_2(\delta\lambda)/A_2$$

where Φ_4 denotes the first four terms on the right-hand side of Eq. (2.4.2) Tables of Y and Z are given in Appendix C and graphs of $Z(\sigma, m_1, 0) \equiv \xi_2$ and $\delta(\sigma, m_1, 0)$ are given in Figs. 5 and 6.

It should be noted that the expression for the pressure in the reflected pulse as given by Eq. (2.4.1) or Eq. (2.4.2) is singular for true-sonic flow speed behind the incident shock, i. e., for $n_2 = q_2$. This singularity masks the formation of secondary shocks and further comments and a modified perturbation theory valid for all flow speeds will be given in the next section.

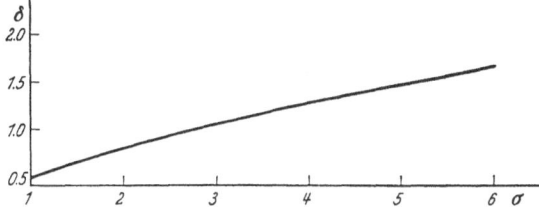

Fig. 6. A graph of the parameter $\delta(\sigma, m_1, 0)$ vs. σ. Since there is only a small total variation with m_1, only the case $m_1 = 0$ is given

2.5 Modified Perturbation Theory

Although the solution for the pressure perturbation directly behind the incident shock, i.e., Eq. (2.3.5) is valid for all flow speeds, the expression for the pressure perturbation in the transient reflected pulse is singular for true-sonic flow speed, i.e., for $u_2 = \omega_2$. In the neighborhood of $u_2 = \omega_2$, the disturbance S tends to accumulate since its speed of propagation is $u_2 - \omega_2$ which is quite small. But, in actuality, as soon as an accumulation begins, $u_2 - \omega_2$ will cease to be small, and the disturbance will be convected through the duct. This difficulty is due to the inadequacies of the linearization, which approximates the characteristics $dx/dt = u_2 - \omega_2$ by the rectilinear and parallel family $x - (u_2 - \omega_2)t$ = constant, but when a more careful examination of the negative characteristics is made, a perturbation theory valid for all flow speeds behind the incident shock may be derived, and the aforementioned singularity actually masks the formation of secondary shocks in the flow. The requisite modifications have been given previously for hydromagnetic shock propagation in channels with only spatially dependent area variations and stationary flow in front of the incident shock wave [15]. The discussion of the present section solves this problem for spatially and time-dependent area variations and includes, as a special case, non stationary flow in front of the incident shock for time-independent area variations.

Since it is the neighborhood of $u_2 = \omega_2$ which is important for this discussion, the governing equations are rewritten with u and ω as dependent variables. Since

$$dw = \frac{(\gamma - 1)\, \omega^2 d\omega}{\omega^2 + (\gamma - 2)\, c^2}$$

Eqs. (1.2.20)—(1.2.21) may be transformed to:

$$\omega_t + u\omega_x + [\omega^2 + (\gamma - 2)c^2]u_x/2\omega + A_x u[\omega^2 +$$
$$+ (\gamma - 2)c^2]/2A\,\omega + A_t[\omega^2 + (\gamma - 2)c^2]/2A\,\omega \tag{2.5.1}$$

$$u_t + u u_x + 2\omega^3 \omega_x/[\omega^2 + (\gamma - 2)c^2] - c^2 s_x/\gamma(\gamma - 1)c_v = 0 . \tag{2.5.2}$$

Subtracting these two equations gives:

$$u_t - \omega_t + (u - \omega)(u_x - \omega_x) + \frac{\omega^2 + (2 - \gamma)c^2}{\omega}\left[\frac{u_x}{2} + \frac{\omega^2 \omega_x}{\omega^2 + (\gamma - 2)c^2}\right]$$

$$= \frac{c^2 s_x}{\gamma(\gamma - 1)c_v} + \frac{u[\omega^2 + (\gamma - 2)c^2]A_x}{2A\,\omega} + \frac{[\omega^2 + (\gamma - 2)c^2]A_t}{2A\,\omega} \tag{2.5.3}$$

which is the equation satisfied along the negative characteristics. The equations valid along the positive characteristics and particle paths are:

$$\frac{u_t}{2} + \frac{\omega^2 \omega_t}{\omega^2 + (\gamma - 2)c^2} + (u + \omega)\left[\frac{u_x}{2} + \frac{\omega^2 \omega_x}{\omega^2 + (\gamma - 2)c^2}\right]$$

$$= \frac{c^2 s_x}{2\gamma(\gamma - 1)c_v} - \frac{u\omega A_x}{2A} - \frac{\omega A_t}{2A} . \tag{2.5.4}$$

$$s_t + u s_x = 0 . \tag{2.5.5}$$

The system of Eqs. (2.5.3)—(2.5.5) is that to be utilized throughout the remainder of this section. For time-independent area variations, this system reduces to that discussed in [15].

Because Eqs. (2.5.4)—(2.5.5) are valid for all flow speeds, they may be linearized directly and lead to the following solutions:

$$R = \frac{\bar{u}_2}{2} + \frac{\omega_2^2 \bar{\omega}_2}{\omega_2^2 + (\gamma - 2)c_2^2} = F[x - (u_2 + \omega_2)t] +$$

$$+ \frac{E[x - u_2 t]}{\omega_2} - \frac{u_2 \omega_2}{2A_2(u_2 + \omega_2)}\bar{A}_2\left[x, \frac{(u_2 + \omega_2)t}{u_2}\right]. \tag{2.5.6}$$

$$T_0 = 2E'[x - u_2 t] . \tag{2.5.7}$$

It is assumed that the physical set-up of the problem is exactly that of section 2.3, and by setting $F \equiv 0$, Eqs. (2.5.6)—(2.5.7) lead directly to the solution for the pressure perturbation behind the shock given by Eq. (2.3.5).

Linearization of Eq. (2.5.3) must be carried out with greater care. It is the terms in $u_2 - \omega_2 \equiv \Omega$ which cause the difficulty and these terms are retained while the remaining terms may be linearized directly. Thus, the resultant equation is:

$$\Omega_t + \Omega\,\Omega_x + [\omega_2^2 + (2-\gamma)\,c_2^2]\,\frac{R_x}{\omega_2} = \frac{c_2^2\bar{s}_{2x}}{\gamma\,(\gamma-1)\,c_v} +$$

$$+ \frac{[\omega_2^2 + (\gamma-2)\,c_2^2]}{2\,A_2\,\omega_2}\,[\bar{A}_{2t} + u_2\bar{A}_{2x}]\;.$$

Substituting from Eqs. (2.5.6)—(2.5.7) gives the following equation valid along the negative characteristics:

$$\Omega_t + \Omega\,\Omega_x = [\omega_2^2 + (\gamma-2)\,c_2^2]\,\frac{E'\,[x - u_2\,t]}{\omega_2^2} +$$

$$+ \frac{u_2\,[\omega_2^2 + (2-\gamma)\,c_2^2]}{2\,A_2\,(u_2 + \omega_2)}\,\bar{A}_{2x}\left[x,\,\frac{(u_2 + \omega_2)\,t}{u_2}\right] + \qquad (2.5.8)$$

$$+ \frac{[\omega_2^2 + (\gamma-2)\,c_2^2]}{2\,A_2\,\omega_2}\,[\bar{A}_{2t}(x,\,t) + u_2\bar{A}_{2x}(x,\,t)]\;.$$

The function E may be expressed in terms of the cross-sectional area perturbation by noting that

$$\frac{\bar{s}_2}{c_v} = \frac{2\,\gamma\,(\gamma-1)\,E\,[x - u_2 t]}{c_2^2} = \frac{\bar{P}_2}{P_2} - \frac{\gamma\,\bar{\varrho}_2}{\varrho_2}\;. \qquad (2.5.9)$$

From the shock conditions, Eq. (2.3.3),

$$\frac{\bar{\varrho}_2}{\varrho_2} = \frac{1}{L_1}\frac{\bar{P}_2}{P_2} - \frac{1}{L_1}\frac{\bar{P}_1}{P_1} + \frac{1}{L_2}\frac{\bar{m}_1}{m_1} + \frac{\bar{\varrho}_1}{\varrho_1}\;, \qquad (2.5.10)$$

where $(L_1,\,L_2)$ refer to the respective coefficients in Eq. (2.3.3). Thus, evaluating Eq. (2.5.9) on the shock, $x = U\,t$, replacing $\dfrac{\bar{P}_2}{P_2}$ by its value obtained from Eq. (2.3.5) and using Eq. (2.5.10), it follows that

$$\frac{E\,[x - u_2 t]}{c_2^2} = \frac{(\gamma - L_1)\,(\tau - 1)}{2\,\gamma\,(\gamma - 1)\,L_1\tau A_2}\,K\,(\sigma,\,m_1,\,n_1)\,\bar{A}_2 \times$$

$$\times \left[\frac{(M_2 + n_2)\,(x - u_2 t)}{M_2},\,\frac{(n_2 + q_2)\,(x - u_2 t)}{n_2 M_2 c_2}\right] +$$

$$+ \frac{(L_1 - \gamma)\,(\tau - 1)}{2\,\gamma\,(\gamma - 1)\,L_1\tau}\,\sum_{i=1}^{5} D_i\,(\sigma,\,m_1,\,n_1)\,\phi_i \times \qquad (2.5.11)$$

$$\times \left[\frac{(M_2 + n_2)\,(x - u_2 t)}{M_2},\,\frac{(x - u_2 t)}{M_2 c_2}\right] +$$

$$+ \frac{1}{L_1}\frac{\phi_4}{2\,(\gamma - 1)} - \frac{1}{L_2}\frac{\phi_5}{2\,(\gamma - 1)} - \frac{\phi_1}{2\,(\gamma - 1)}\;,$$

where the arguments of the last three terms are the same as in the summation term. For convenience, let Eq. (2.5.11) be rewritten with a new

26   2. Shock Propagation in Non-Uniform Ducts

definition of coefficients as:

$$\frac{E\,[x-u_2 t]}{c_2^2} = \beta_1 \frac{\bar{A}_2}{A_2}\left[\frac{(M_2+n_2)\,(x-u_2 t)}{M_2}, \frac{(n_2+q_2)\,(x-u_2 t)}{M_2 n_2 c_2}\right]+$$

$$+ \sum_{i=1}^{5} \delta_i(\sigma, m_1, n_1)\,\phi_i\left[\frac{(M_2+n_2)\,(x-u_2 t)}{M_2}, \frac{(x-u_2 t)}{M_2 c_2}\right]$$

so that

$$\frac{E'\,[x-u_2 t]}{c_2^2} = \beta_1 \frac{(M_2+n_2)\,\bar{A}_{2,1}}{M_2 A_2} + \beta_1 \frac{(n_2+q_2)}{M_2 n_2 c_2}\frac{\bar{A}_{2,2}}{A_2} +$$

$$+ \sum_{i=1}^{5} \delta_i\,[\phi_{i,1}(M_2+n_2)/M_2 + \phi_{i,2}/M_2 c_2] \tag{2.5.12}$$

where the subscripts after the commas refer to derivatives with respect
to the first or second argument of the pertinent function. Substitution
of Eq. (2.5.12) into Eq. (2.5.8) gives the final differential equation, viz.,

$$\Omega_t + \Omega\,\Omega_x = \frac{(q_2^2+\gamma-2)\,\beta_1 c_2^2 (M_2+n_2)\,\bar{A}_{2,1}}{q_2^2 M_2 A_2} +$$

$$+ (n_2+q_2)\frac{(q_2^2+\gamma-2)\,\beta_1 c_2 \bar{A}_{2,2}}{q_2^2 M_2 n_2 A_2} + \frac{(q_2^2+\gamma-2)\,c_2^2 (M_2+n_2)}{q_2^2 M_2}\sum_{i=1}^{5}\delta_i\,\phi_{i,1} +$$

$$+ \frac{(q_2^2+\gamma-2)\,c_2}{q_2^2 M_2}\sum_{i=1}^{5}\delta_i\,\phi_{i,2} + \frac{(q_2^2+2-\gamma)n_2 c_2^2}{2 A_2(n_2+q_2)}\bar{A}_{2,1}\,[x,(n_2+q_2)t/n_2] +$$

$$+ \frac{(q_2^2+\gamma-2)\,c_2}{2 A_2 q_2}\,[u_2\bar{A}_{2,1}(x,t)+\bar{A}_{2,2}(x,t)] \tag{2.5.13}$$

where the arguments have been omitted in the functions arising from
Eq. (2.5.12). For u_2 not close to ω_2, Eq. (2.5.13) may be solved directly
and the solution so obtained agrees with Eq. (1.2.31) after a redefinition
of the arbitrary function G, and this leads to the solution presented in
section 2.4.

For u_2 near ω_2, it is a first order non-linear partial differential equation
with *prescribed* inhomogeneous term and is to be integrated along the
negative characteristics. The initial condition at the shock is easily
determined. Since

$$\frac{\bar{\omega}_2}{\omega_2} = \frac{[q_2^2(2+\eta^*)-\eta^*]\,\bar{c}_2}{2 q_2^2 c_2}, \quad \frac{\bar{c}_2}{c_2} = \frac{\bar{c}_1}{c_1} + \frac{\bar{\tau}}{2\tau} - \frac{\bar{\sigma}}{2\sigma}$$

and $\bar{\sigma}/\sigma$ and $\bar{\tau}/\tau$ are known on the shock, viz., Eqs. (2.3.5) and (2.5.10),

$$\frac{\bar{\omega}_2}{\omega_2} = \left[\frac{q_2^2(2+\eta^*)-\eta^*}{2 q_2^2}\right]\left[\phi_3\,(Ut,t) - \frac{\phi_5\,(Ut,t)}{2 L_2} + \frac{(1-L_1)\,\phi_4(Ut,t)}{2 L_1} +\right.$$

$$\left.+ \frac{(L_1-1)}{2 L_1}\frac{(\tau-1)}{\tau}\sum_{i=1}^{5} D_i(\sigma,m_1,n_1)\,\phi_i(Ut,t)\right]+$$

$$+ \frac{[q_2^2(2+\eta^*)-\eta^*]\,(1-L_1)\,(\tau-1)\,K\,(\sigma,m_1,n_1)\,\bar{A}_2\,[Ut,(n_2+q_2)t/n_2]}{4 q_2^2 L_1 \tau A_2} \tag{2.5.14}$$

on the shock. Since $\Omega = u - \omega = \bar{u}_2 - \bar{\omega}_2 + u_2 - \omega_2$, with $\bar{\omega}_2$ given by Eq. (2.5.14) and $\bar{u}_2$ by Eq. (2.3.5), the initial condition may be written as:

$$\Omega = u_2 - \omega_2 + \sum_{i=1}^{5} Q_i \phi_i (Ut, t) - Q_6 \frac{\bar{A}_2}{A_2} [Ut, (n_2 + q_2) t/n_2] \qquad (2.5.15)$$

with appropriate coefficients $Q_j, j = 1, 2, 3, 4, 5, 6$. Thus the problem is reduced to solving Eq. (2.5.13) with the initial condition (2.5.15) given at the shock. If the area perturbation is approximated by a few terms of a Taylor series, Eq. (2.5.13) may be solved exactly to that order of approximation and the initial condition Eq. (2.5.15) applied. For further details, including a determination of the point of secondary shock formation, the previous paper [15] may be consulted. For time-independent area variations, the discussion of this section gives the modified perturbation theory for the propagation of a non-uniform shock into a moving fluid.

2.6 Criterion for the Particle Velocity to be Unaffected by Entropy Perturbations

It was shown in Chapter 1 that the addition of an entropy perturbation introduced a non-homogeneous term in an otherwise homogeneous system of perturbation equations, and the entropy perturbation could be determined directly. This meant that problem could be solved by first considering the homogeneous system (isentropic perturbed flow) and then adding particular solutions to the complete system (non-isentropic s erturbed flow). For two conventional gas dynamic flows of interest, viz., an initially uniform or centered simple wave flow, it was found previously [2], [5], [6] that the addition of an entropy perturbation affected the pound speed but not the particle velocity, i.e., there was a *particular* solution with the particle velocity perturbation equal to zero. A general discussion of this phenomenon, including necessary and sufficient conditions for it to occur, was given [5], [6].

In the present theory, it has been seen that the addition of an entropy perturbation did not affect the particle velocity in an initially uniform flow [Eqs. (1.2.30)−(1.2.32)], which is a direct consequence of the result that the non-isentropic perturbation of an initially uniform flow must reduce to the solution of the corresponding problem in conventional gas dynamics in the limit of vanishing magnetic field, but, in contrast to ordinary gas dynamics, this result *did not* obtain for a centered simple wave flow [30]. Simple wave perturbations will be discussed in detail in Chapter 5.

Consequently, it is of interest to examine the question in greater detail [31]. In order to be able to use the perturbation equations which have been derived, the discussion is limited to the class of flows for which

$B/c^{2/(\gamma-1)}$ is constant throughout the flow. It is convenient to use (u, ω, s) as the dependent variables, so that the pertinent equations are

$$\omega_t + u\omega_x + \frac{[\omega^2 + (\gamma - 2)\,c^2]}{2\omega}\,u_x = 0 \tag{2.6.1}$$

$$u_t + u\,u_x + \frac{2\omega^3\,\omega_x}{[\omega^2 + (\gamma - 2)\,c^2]} = \frac{c^2 s_x}{\gamma(\gamma - 1)c_v} \tag{2.6.2}$$

$$s_t + u s_x = 0 \,. \tag{2.6.3}$$

When this system of equations is linearized in the neighborhood of a known isentropic flow, denoted by the subscript zero, a system of linear equations for the terms of first order, denoted by the subscript one, is obtained exactly as in Chapter 1. Since for the problem under consideration, it is equivalent to look for solutions of the linear system with $u_1 = 0$, the pertinent equations are

$$\omega_{1t} + u_0\omega_{1x} + \frac{[\omega_0^2 + (\gamma - 2)\,c_0^2]\,u_{0x}\omega_1}{2\omega_0^2} = 0 \tag{2.6.4}$$

$$(\omega_0\omega_1)_x = \frac{[\omega_0^2 + (\gamma - 2)\,c_0^2]}{(\gamma - 1)\,\omega_0^2}\,\frac{\varrho_0 c_0^2\,\Gamma'\,(\varPsi_0)}{2\,\gamma c_v} \tag{2.6.5}$$

$$d\varPsi_0 = \varrho_0\,[dx - u_0 dt] \tag{2.6.6}$$

where $\varPsi_0 =$ constant defines the particle paths. Consequently, the problem is reduced to expressing the compatibility of Eqs. (2.6.4) and (2.6.5), and the conditions on the functions $u_0(x, t)$, $c_0(x, t)$ and $\omega_0(x, t)$ which allow this compatibility will be determined.

The characteristics of Eq. (2.6.4) are

$$\frac{dt}{1} = \frac{dx}{u_0} = \frac{d\omega_1}{-\left[\dfrac{\omega_0^2 + (\gamma - 2)\,c_0^2}{2\omega_0^2}\right]u_{0x}\omega_1} \tag{2.6.7}$$

One first integral of Eq. (2.6.7) is $\varPsi_0 =$ constant, i.e., the particle paths. Since $\omega_1 = \omega_0$ is a *particular* solution of Eq. (2.6.4), the following method of solution is suggested. From Eq. (2.6.7)

$$\frac{\omega_{0t}dt}{\omega_{0t}} = \frac{\omega_{0x}dx}{u_0\omega_{0x}} = \frac{\omega_{0t}dt + \omega_{0x}dx}{\omega_{0t} + u_0\omega_{0x}}$$

$$= \frac{d\omega_0}{-\left[\dfrac{\omega_0^2 + (\gamma - 2)\,c_0^2}{2\omega_0}\right]u_{0x}} = \frac{d\omega_1}{-\left[\dfrac{\omega_0^2 + (\gamma - 2)\,c_0^2}{2\omega_0^2}\right]u_{0x}\omega_1}$$

which shows clearly that $\omega_1/\omega_0 =$ constant is a first integral. Consequently the solution of Eq. (2.6.4) may be written

$$\frac{\omega_1}{\omega_0} = f(\varPsi_0) \tag{2.6.8}$$

where f is an arbitrary differentiable function.

Since Eq. (2.6.4) is obtained by linearizing the continuity equation, which is the same in the magnetic or non-magnetic case, it follows that [5], [6]

$$\frac{c_1}{c_0} = g(\Psi_0) \tag{2.6.9}$$

where g is an arbitrary differentiable function.

Since it follows from Eqs. (2.6.8) and (2.6.9) that

$$\frac{\omega_1}{c_1} = \frac{\omega_0}{c_0} = \frac{f(\Psi_0)}{g(\Psi_0)} = \frac{\omega_0 [\omega_0^2 + (\gamma - 2) c_0^2]}{(\gamma - 1) c_0 \omega_0^2},$$

Eq. (2.6.5) may be written as

$$(\omega_0 \omega_1)_x = \varrho_0 c_0^2 \left[\frac{\Gamma'(\Psi_0) f(\Psi_0)}{2 \gamma c_v g(\Psi_0)} \right]. \tag{2.6.10}$$

Substituting Eq. (2.6.8) into Eq. (2.6.10) gives

$$\frac{\omega_{0x}}{\varrho_0 \omega_0} = \frac{c_0^2}{\omega_0^2} \left[\frac{\Gamma'(\Psi_0)}{4 \gamma c_v g(\Psi_0)} \right] - \frac{f'(\Psi_0)}{2 f(\Psi_0)} \tag{2.6.11}$$

Writing $\varrho_0 = r_2 c_0^{\frac{2}{\gamma-1}}$ and absorbing the constant r_2 in the arbitrary functions, replacing $\frac{\omega_{0x}}{\omega_0} = \frac{c_{0x} f(\Psi_0)}{c_0 g(\Psi_0)}$ and taking the material derivative in Eq. (2.6.11), the following condition is obtained

$$\frac{D}{Dt} \left[\frac{c_{0x}}{c_0^\theta} \right] = \frac{\Gamma'(\Psi_0)}{4 \gamma c_v f(\Psi_0)} \frac{D}{Dt} \left[\frac{c_0^2}{\omega_0^2} \right]. \tag{2.6.12}$$

When the differentiation on the right-hand side of Eq. (2.6.12) is carried out and c_{0xt} is replaced by its equivalent obtained by differentiating the continuity equation with respect to x, the following result is obtained

$$\frac{u_{0xx}}{c_0^{\frac{2}{\gamma-1}}} = - \frac{\Gamma'(\Psi_0)}{2 (\gamma - 1) \gamma c_v f(\Psi_0)} \frac{D}{Dt} \left[\frac{c_0^2}{\omega_0^2} \right]. \tag{2.6.13}$$

For the non-magnetic case, the right-hand side vanishes and the condition is

$$u_{0xx} = 0$$

so that

In the non-magnetic case, a necessary and sufficient condition for the particle velocity to be unaffected by the addition of an entropy perturbation is that $u_0(x, t)$ be a linear function of x.

An investigation of all solutions of this form, i.e., $u_0 = a_1(t) x + b_1(t)$, has been carried out [5].

When the indicated differentiation in Eq. (2.6.13) is carried out and the material derivative is again taken to eliminate the functions of Ψ_0, the condition may be written as

$$\frac{D}{Dt} \left[\frac{\frac{u_{0xx} \omega_0^2}{2}}{u_{0x} c_0^2 c_0^{\frac{2}{\gamma-1}}} \right] = 0 \tag{2.6.14}$$

which may be put into the following form

$$\frac{D}{Dt}\left[\frac{u_{0xx}}{u_{0x}}\right] + \left[1 + \frac{(\gamma - 2)(\omega_0^2 - c_0^2)}{\omega_0^2}\right] u_{0xx} = 0 \,. \qquad (2.6.15)$$

Thus, the final result

In the magnetic case, a necessary and sufficient condition for the particle velocity to be unaffected by the addition of an entropy perturbation is that (u_0, c_0, ω_0) satisfy (2.6.14) or (2.6.15).

It should be noted that $u_{0xx} = 0$ is a *sufficient* condition.

Chapter 3

The Piston-Driven Shock Wave

3.1 Arbitrary Area Variations

In Chapter 2, the effects of cross-sectional area variations on the motion of an initially uniform shock wave were determined with no reference to the method of generation of the shock. It is the purpose of the present chapter to consider the additional effects of waves reflected back from a piston, the initially uniform motion of which generated the shock. The solutions contain, as special cases, the solutions for the corresponding problems in conventional gas dynamic flows. The effects of small non-uniformities in piston motion on the propagation of an initially uniform shock wave in a uniform duct was determined in [5] and [6]; in ducts with linear area variations (cylindrical shock wave) in [6] and [12]; in ducts with quadratic area variations (spherical shock wave) in [17]; in ducts with arbitrary area variations in [18] and [19].

Taking the usual model, it is assumed that the piston, initially at rest, is moved impulsively with initially constant speed into the fluid. For simplicity, it is assumed that the area variations are time-independent, the fluid in front of the shock is at rest and the piston starts from $x = 0$ at $t = 0$. The effects of the area variations and small non-uniformities of the piston motion may be determined from the general solution for the non-isentropic perturbation of an initially uniform flow as given by Eqs. (1.2.30)—(1.2.32). For ready reference, this solution is rewritten here specialized to the case of time-independent area variations, viz.,

$$R = F\left[x - (u_2 + \omega_2)t\right] + \frac{E\left[x - u_2 t\right]}{\omega_2} - \frac{u_2 \omega_2 \overline{A}_2[x]}{2(u_2 + \omega_2)A_2} \qquad (3.1.1)$$

$$S = G\left[x - (u_2 - \omega_2)t\right] + \frac{E\left[x - u_2 t\right]}{\omega_2} - \frac{u_2 \omega_2 \overline{A}_2[x]}{2(u_2 - \omega_2)A_2} \qquad (3.1.2)$$

$$T_0 = 2E'\left[x - u_2 t\right] \,. \qquad (3.1.3)$$

When the shock propagates through the non-uniform duct, it is perturbed and the initially isentropic flow behind the shock becomes weakly non-isentropic. The compressive and rarefactive wavelets thus generated propagate with true sonic speed with respect to the flow, are reflected at the piston and return to modify the motion of the shock. These are again reflected at the shock and the pattern repeats itself indefinitely. In addition, further perturbations are generated by small non-uniformities of the piston motion. Thus there is an extremely complicated interaction pattern created behind the shock, but an expression for the average pressure perturbation behind the shock may be obtained rather simply from the present theory.

From the shock conditions, it follows that R, S and $\bar{s}_2$ may be expressed in terms of the pressure perturbation at the shock. This gives:

$$R = N_1 \bar{\tau}/\tau; \quad S = N_2 \bar{\tau}/\tau; \quad \bar{s}_2/2\,\gamma\,(\gamma - 1)\,c_v = N_3 \bar{\tau}/\tau \quad (3.1.4)$$

at the shock, $x = Ut$, where (N_1, N_2, N_3) are the appropriate coefficients. Using the subsidiary conditions Eq. (3.1.4), the arbitrary functions in Eqs. (3.1.1)—(3.1.2) may be determined in terms of $\bar{\tau}/\tau$. Since it is assumed that the piston velocity perturbation is prescribed, say $\bar{u}_{2p}(t)$, the following functional equation is obtained by evaluating $\bar{u}_2 = R - S$ at the piston curve, $x = u_2 t$.

$$\frac{1}{u_2}\left[N_1 - \frac{c_2 N_3}{q_2}\right]\frac{\bar{\tau}\,[q_2 t/(q_2 - M_2)]}{\tau} + \frac{1}{u_2}\left[N_3 \frac{c_2}{q_2} - N_2\right]\frac{\bar{\tau}\,[q_2 t/(M_2 + q_2)]}{\tau}$$

$$= -\frac{q_2}{2\,(n_2 + q_2)\,A_2}\,\bar{A}_2\left[\frac{(M_2 + n_2)\,\omega_2 t}{q_2 - M_2}\right] +$$

$$+ \frac{q_2}{2\,A_2\,(n_2 - q_2)}\,\bar{A}_2\left[\frac{(M_2 + n_2)\,\omega_2 t}{M_2 + q_2}\right] + \frac{q_2}{2\,A_2\,(n_2 + q_2)}\,\bar{A}_2\,[u_2 t] - $$

$$-\frac{q_2}{2\,A_2\,(n_2 - q_2)}\,\bar{A}_2\,[u_2 t] + \bar{u}_{2p}\,(t)/u_2 \equiv J\,(t)\,.$$

(3.1.5)

The functional equation (3.1.5) is a special case of the theorem proved in [6], p. 578. Thus the solution may be written down immediately.

$$\frac{\bar{\tau}\,[q_2 t/(q_2 - M_2)]}{\tau}$$

$$= \sum_{\nu=0}^{\infty} \frac{[N_3 c_2/u_2 q_2 - N_2/u_2]^\nu\,(-1)^\nu}{[N_1/u_2 - c_2 N_3/u_2 q_2]^{\nu+1}}\,J\,[t\{(q_2 - M_2)/(q_2 + M_2)\}^\nu] \equiv \quad (3.1.6)$$

$$\equiv \sum_{\nu=0}^{\infty} (-1)^\nu\,\Omega_\nu\,(\sigma,\,m_1,\,\nu)\,J\,[t\{(q_2 - M_2)/(q_2 + M_2)\}^\nu]$$

where

$$\Omega_\nu = \frac{\Gamma_1^\nu}{\Gamma_2^{\nu+1}}$$

and

$$\Gamma_1 = \frac{(1-q_2^2)}{2(\gamma-1)n_2 q_2} \times$$

$$\times \left[1 - \frac{(\theta-\sigma)\,[\theta\sigma - 1 - \gamma m_1^2(1-\sigma)^3/2]}{(\theta^2-1)\sigma + m_1^2\gamma\sigma[3\theta - 1 - 6\theta\sigma + 3(\theta+1)\sigma^2 - 2\sigma^3]/2}\right] -$$

$$- \frac{1}{2\gamma n_2 q_2} +$$

$$+ \frac{\dfrac{2(\theta+1)\theta}{\gamma} + \dfrac{2(\theta+1)(\theta-2)\sigma}{\gamma} + m_1^2[\theta(\theta+1) + (\theta+1)(\theta-2)\sigma + (\theta-3)(2\theta-1)\sigma^2 + (3-\theta)\sigma^3]}{\dfrac{4(\sigma-1)\{2(\theta+1)/\gamma + m_1^2[1+\theta+\sigma(\theta-3)]\}}{(\theta^2-1)\sigma + m_1^2\gamma\sigma[3\theta-1-6\theta\sigma+3(\theta+1)\sigma^2-2\sigma^3]/2}}{\theta\sigma-1-\gamma m_1^2(1-\sigma)^3/2}$$

$$\Gamma_2 = \frac{1}{2\gamma n_2 q_2} + \frac{(q_2^2-1)}{2(\gamma-1)n_2 q_2} \times$$

$$\times \left[1 - \frac{(\theta-\sigma)\,[\theta\sigma - 1 - \gamma m_1^2(1-\sigma)^3/2]}{(\theta^2-1)\sigma + m_1^2\gamma\sigma[3\theta-1-6\theta\sigma+3(\theta+1)\sigma^2-2\sigma^3]/2}\right] +$$

$$+ \frac{\dfrac{2(\theta+1)\theta}{\gamma} + \dfrac{2(\theta+1)(\theta-2)\sigma}{\gamma} + m_1^2[\theta(\theta+1) + (\theta+1)(\theta-2)\sigma + (\theta-3)(2\theta-1)\sigma^2 + (3-\theta)\sigma^3]}{\dfrac{4(\sigma-1)\{2(\theta+1)/\gamma + m_1^2[1+\theta+\sigma(\theta-3)]\}}{(\theta^2-1)\sigma + m_1^2\gamma\sigma[3\theta-1-6\theta\sigma+3(\theta+1)\sigma^2-2\sigma^3]/2}}{\theta\sigma-1-\gamma m_1^2(1-\sigma)^3/2}$$

For $m_1 = 0$, Eq. (3.1.6) reduces to the result given in [19] for the corresponding problem for non-conducting fluids. The series, Eq. (3.1.6), converges rapidly, and a few terms suffice for a reasonable approximation. An alternative approach is to approximate the area perturbation by a Taylor series, and then the resultant functional equation may be solved term by term.

3.2 The Piston-Driven Cylindrical Shock Wave

The cylindrical shock wave may be discussed as a special case of shock propagation in a non-uniform duct by choosing a wedge-shaped duct with cross-sectional area proportional to the distance from the center of the shock front, i.e., a linear area variation. The motion of a piston-driven cylindrical shock wave may be determined as a special case of the results of section 3.1, but it is easier to approach this problem directly since its solution may be given in finite form. Choosing $\bar{A}_2(x) = \mathbb{E}\,x$, the equations corresponding to Eqs. (3.1.1)—(3.1.2) are

$$R = F[x - (u_2 + \omega_2)t] + \frac{E[x - u_2 t]}{\omega_2} - \frac{u_2 \omega_2 \mathbb{E}\,x}{2A_2(u_2 + \omega_2)} \qquad (3.2.1)$$

$$S = G[x - (u_2 - \omega_2)t] + \frac{E[x - u_2 t]}{\omega_2} - \frac{u_2 \omega_2 \mathbb{E}\,x}{2A_2(u_2 - \omega_2)}. \qquad (3.2.2)$$

The solution is obtained exactly as in section 3.1. The subsidiary conditions are again Eq. (3.1.4), and this allows the determination of F and G in terms of $\bar{\tau}/\tau$. With the assumption that there is no perturbation of the piston, the following functional equation is obtained

$$\left[N_1 - \frac{c_2 N_3}{q_2}\right] \frac{\bar{\tau} [q_2 t/(q_2 - M_2)]}{\tau} + \left[\frac{c_2}{q_2} N_3 - N_2\right] \frac{\bar{\tau} [q_2 t/(M_2 + q_2)]}{\tau}$$
$$= \frac{(M_2 n_2 - q_2^2) u_2 q_2 \omega_2 (M_2 + n_2) \in t}{A_2 (n_2^2 - q_2^2) (M_2^2 - q_2^2)} - \frac{u_2^2 q_2^2 \in t}{A_2 (n_2^2 - q_2^2)} . \tag{3.2.3}$$

A linear function of t satisfies Eq. (3.2.3), so that the solution may be written as

$$\frac{\bar{\tau}}{\tau} = \frac{\left[\dfrac{u_2 q_2 \omega_2 (M_2 n_2 - q_2^2) (M_2 + n_2)}{(M_2^2 - q_2^2)(n_2^2 - q_2^2)} - \dfrac{u_2^2 q_2^2}{(n_2^2 - q_2^2)} \right] \dfrac{\in t}{A_2}}{\dfrac{q_2}{(q_2 - M_2)} \left[N_1 - \dfrac{c_2 N_3}{q_2} \right] + \dfrac{q_2}{M_2 + q_2} \left| \dfrac{c_2 N_3}{q_2} - N_2 \right|} \tag{3.2.4}$$

Eq. (3.2.4) may be put into the form

$$\bar{P}_2 = - K_c(\sigma, m_1) (P_2 - P_1) \bar{A}_2 / A_2 \tag{3.2.5}$$

where $K_c = Q_7 / Q_8$ and

$$Q_7 = n_2 q_2^2 \left[\frac{\theta \sigma - 1 - \gamma m_1^2 (1 - \sigma)^3/2}{\theta + 1 + \gamma m_1^2 (\sigma - 1)^2/2} \right] \left[\frac{n_2 (M_2^2 - q_2^2)}{M_2 + n_2} + q_2^2 - n_2 M_2 \right]$$

$$Q_8 = - M_2 (\sigma - 1) (n_2^2 - q_2^2)/\gamma + M_2 (n_2^2 - q_2^2) (\sigma - 1) (1 - q_2^2) \times$$

$$\times \left\{ 1 - \frac{(\theta - \sigma) [\theta \sigma - 1 - \gamma m_1^2 (1 - \sigma)^3/2]}{(\theta^3 - 1)\sigma + m_1^3 \gamma \sigma [3\theta - 1 - 6\theta\sigma + 3(\theta + 1)\sigma^2 - 2\sigma^3]/2} \right\} +$$

$$+ n_2^2 q_2^2 (q_2^2 - n_2^2) \times$$

$$\times \left\{ \frac{\dfrac{2(\theta+1)\theta/\gamma + 2(\theta+1)(\theta-2)\sigma/\gamma + m_1^2[\theta(\theta+1) + (\theta+1)(\theta-2)\sigma + (\theta-3)(2\theta-1)\sigma^2 + (3-\theta)\sigma^3]}{2\{2(\theta+1)/\gamma + m_1^2[1 + \theta + \sigma(\theta - 3)]\}}}{\dfrac{(\theta^2 - 1)\sigma + m_1^2 \gamma \sigma [3\theta - 1 - 6\theta\sigma + 3(\theta + 1)\sigma^2 - 2\sigma^3]/2}{\theta \sigma - 1 - \gamma m_1^2 (1 - \sigma)^3/2}} \right\}$$

Tables of the parameter $K_c(\sigma, m_1)$ are given in Appendix E. For $m_1 = 0$, K_c decreases monotonically with increasing σ and agrees exactly with the corresponding parameter in the non-conducting case [12]. For all m_1, $\lim_{\sigma \to 1+} K_c = 0.5$ and $\lim_{\sigma \to 6-} K_c = 0.259259$ for $\gamma = 7/5$. For an arbitrary m_1, the behavior is more complicated. For $m_1 \geq 3$, K_c decreases monotonically to a minimum, increases monotonically to a maximum and then decreases monotonically to 0.259259. For increasing m_1, the points of these extremal values diverge. Since K_c exhibits a monotonic decrease for $m_1 = 1$ or 2, it is conjectured that the two aforementioned extremal values come together, and this confluence leads to the monotonic behavior. For fixed σ, K_c decreases monotonically with increasing m_1.

3.3 The Piston-Driven Spherical Shock Wave

The spherical shock wave may be discussed as a special case of shock propagation in a non-uniform duct by choosing a conical channel with area proportional to the square of the distance from the center. Thus, choosing $\bar{A}_2(x) = \in x^2$, the equations corresponding to Eq. (3.1.1) — (3.1.2) are

$$R = F\left[x - (u_2 + \omega_2)t\right] + \frac{E\left[x - u_2 t\right]}{\omega_2} - \frac{u_2 q_2 \in x^2}{2A_2(n_2 + q_2)} \qquad (3.3.1)$$

$$S = G\left[x - (u_2 - \omega_2)t\right] + \frac{E\left[x - u_2 t\right]}{\omega_2} - \frac{u_2 q_2 \in x^2}{2A_2(n_2 - q_2)} \qquad (3.3.2)$$

The subsidiary conditions are again Eq. (3.1.4), and this allows the determination of F and G in terms of $\bar{\tau}/\tau$. With the assumption that there is no perturbation of the piston, the following functional equation is obtained:

$$\left[N_1 - \frac{c_2 N_3}{q_2}\right]\frac{\bar{\tau}\left[q_2 t/(q_2 - M_2)\right]}{\tau} + \left[\frac{c_2 N_3}{q_2} - N_2\right]\frac{\bar{\tau}\left[q_2 t/(M_2 + q_2)\right]}{\tau}$$

$$= -\frac{u_2^3 q_2^2 \in t^2}{A_2(n_2^2 - q_2^2)} - \frac{u_2 q_2^2 \omega_2^2 (M_2 + n_2)^2 (2n_2 M_2 - M_2^2 - q_2^2) \in t^2}{A_2(n_2^2 - q_2^2)(M_2^2 - q_2^2)^2} \qquad (3.3.3)$$

A quadratic function of t satisfies Eq. (3.3.3), so the solution may be written as

$$\frac{\bar{\tau}}{\tau} = \frac{\left[\dfrac{u_2 \omega_2^2 (M_2 + n_2)^2 (M_2^2 - 2n_2 M_2 + q_2^2)}{A_2(n_2^2 - q_2^2)(M_2^2 - q_2^2)^2} - \dfrac{u_2^3}{A_2(n_2^2 - q_2^2)}\right]\in t^2}{\left[N_1 - \dfrac{c_2 N_3}{q_2}\right]\dfrac{1}{(q_2 - M_2)^2} + \left[\dfrac{c_2 N_3}{q_2} - N_2\right]\dfrac{1}{(M_2 + q_2)^2}} \qquad (3.3.4)$$

Eq. (3.3.4) may be put into the form

$$\bar{P}_2 = -K_s(\sigma, m_1)(P_2 - P_1)\bar{A}_2/A_2 \qquad (3.3.5)$$

where $K_s = Q_9/Q_{10}$ and

$$-Q_9 = \left[\frac{n_2 q_2^2 (M_2^2 - 2n_2 M_2 + q_2^2)}{n_2^2 - q_2^2} - \right.$$

$$\left. - \frac{n_2^3 (M_2^2 - q_2^2)^2}{(n_2^2 - q_2^2)(M_2 + n_2)^2}\right]\left[\frac{\theta\sigma - 1 - \gamma m_1^2(1-\sigma)^3/2}{\theta + 1 + \gamma m_1^2(\sigma - 1)^2/2}\right]$$

$$Q_{10} = 2M_2(\sigma - 1)/\gamma + \frac{2M_2(q_2^2 - 1)(\sigma - 1)}{\gamma - 1} \times$$

$$\times \left[1 - \frac{(\theta - \sigma)[\theta\sigma - 1 - \gamma m_1^2(1-\sigma)^3/2]}{(\theta^2 - 1)\sigma + m_1^2\gamma\sigma[3\theta - 1 - 6\theta\sigma + 3(\theta + 1)\sigma^2 - 2\sigma^3]/2}\right] + n_2(M_2^2 + q_2^2) \times$$

$$\times \left[\frac{2(\theta+1)\theta/\gamma + 2(\theta+1)(\theta-2)\sigma/\gamma + m_1^2[\theta(\theta+1) + (\theta+1)(\theta-2)\sigma + (\theta-3)(2\theta-1)\sigma^2 + (3-\theta)\sigma^3]}{2\{2(\theta+1)/\gamma + m_1^2[1 + \theta + \sigma(\theta-3)]\}}\right]$$
$$\left[\frac{(\theta^2 - 1)\sigma + m_1^2\gamma\sigma[3\theta - 1 - 6\theta\sigma + 3(\theta + 1)\sigma^2 - 2\sigma^3]/2}{\theta\sigma - 1 - \gamma m_1^2(1-\sigma)^3/2}\right]$$

Tables of the parameter $K_s(\sigma, m_1)$ are given in Appendix E. For $m_1 = 0$, K_s decreases monotonically and agrees exactly with the corresponding parameter in the non-conducting case [17]. For all m_1, $\lim\limits_{\sigma \to 1+} K_s = 0.5$ and $\lim\limits_{\sigma \to 6-} K_s = 0.361111$ for $\gamma = 7/5$. For any $m_1 \neq 0$, the monotonicity is lost, but the curves are concave upward with curves for greater m_1 lying beneath those for lesser m_1, i.e., for fixed σ, K_s decreases monotonically with increasing m_1.

3.4 The Integrated Shock Strength-Area Relation

A relation of the form

$$\overline{P}_2 = -K(\sigma, m_1)\,(P_2 - P_1)\,\frac{\overline{A}_2}{A_2}$$

may be interpreted as a first order differential relation between area and shock strength, viz.,

$$\frac{dA}{A} + \frac{d\tau}{(\tau - 1)K} = 0 \tag{3.4.1}$$

and the solution may be written as

$$A = \nu \exp\left[-\int \frac{d\tau}{(\tau - 1)K}\right] \equiv \nu\,H(\sigma, m_1) \tag{3.4.2}$$

where

$$-\log_e H(\sigma, m_1) \tag{3.4.3}$$
$$= \int \left[\frac{\theta^2 - 1 + m_1^2\,\gamma\,[3\theta - 1 - 6\theta\sigma + 3(1 + \theta)\sigma^2 - 2\sigma^3]/2}{(\theta - \sigma)\,(\sigma - 1)\,[\theta + 1 + \gamma m_1^2(\sigma - 1)^2/2]\,K(\sigma, m_1)}\right] d\sigma$$

and ν is a constant of integration. It is more convenient to give the integral in terms of σ since K is known in terms of σ.

CHISNELL [9] integrated the shock strength-area relation in *closed form* in the non-magnetic case and showed that the result could be used to give an approximate description of the motion of a shock in terms of the area of the duct. By suitable choices of the cross-sectional area distribution, a description of converging cylindrical (linear area variation) and spherical (quadratic area variation) shock waves was given. These results were checked by comparison with previous similarity solutions, valid in the neighborhood of the points of collapse of the shock, and the comparison showed the remarkable accuracy of CHISNELL's work.

CHISNELL's work was based on the motion of an initially uniform shock wave with no specification of the method of generation thereof. It was shown by GUNDERSEN that these results could be extended to piston-driven shocks, where waves reflected back from a piston return to modify the motion of the shock. Closed-form shock strength-area relationships were obtained for cylindrical shock waves [12], i.e., a linear area variation, and for spherical shock waves [17], i.e., a quadratic area variation.

3*

In contrast to plane shock waves, converging cylindrical and spherical shock waves are unstable and ultimately become strong. As the shock converges, its strength increases and ultimately becomes singular at the center. Near the singular point, Eq. (3.4.1) shows that A is proportional to τ^{-1/K^*}, where K^* is the asymptotic limit of K, so that the shock strengths of converging cylindrical and spherical shocks near the singular point are proportional, respectively, to D^{-K^*} and D^{-2K^*}, where D is the distance of the shock from its axis or point of symmetry. For $\gamma = 7/5$, CHISNELL showed that the strengths were proportional to $D^{-0.394141}$ and $D^{-.788282}$. It was shown by GUNDERSEN that the strengths of converging cylindrical and spherical piston-driven shocks were proportional to $D^{-0.259259}$ and $D^{-0.361111}$, respectively.

For monatomic fluids, it was shown by GUNDERSEN [21] that CHISNELL's work could be extended to the magnetic case, and a theory of converging cylindrical and spherical magnetohydrodynamic shock waves was given. This theory, which included CHISNELL's work as a special case, was extended later to an arbitrary value of the adiabatic index [4]. In these papers, it was shown that

Near the axis or point of symmetry, the strengths of converging cylindrical and spherical magnetohydrodynamic shock waves are independent of the applied field and given by conventional gas dynamic theory.

It is the purpose of the present section to point out that the theory of [12] and [17] may be extended to the magnetic case and in a form which includes the conventional gas dynamic results as a special case. The integrated shock strength-area relation must be obtained numerically, and Eq. (3.4.3) is the basic relation. Numerical results are given in Appendix E, and the tables refer to the evaluation of

$$F(\sigma, m_1) = -\log_e H(\sigma, m_1)$$
$$= \int_{1.01}^{\sigma} \left[\frac{\theta^2 - 1 + m_1^2 \gamma [3\theta - 1 - 6\theta x + 3(1 + \theta) x^2 - 2x^3]/2}{(\theta - x)(x - 1)[\theta + 1 + \gamma m_1^2 (x - 1)^2/2] K(x, m_1)} \right] dx \qquad (3.4.4)$$

for $\gamma = 7/5$. The piston-driven cylindrical shock wave is obtained by letting $K = K_c$ and the resultant function is called $F_c(\sigma, m_1)$. The piston-driven spherical shock wave is obtained by letting $K = K_s$ and the resultant function is called $F_s(\sigma, m_1)$. Data are presented for several values of m_1, the measure of the applied field, including the conventional gas dynamic case which corresponds to $m_1 = 0$. For convenience of reference, a table of values of the function $F(\sigma, m_1)$ for the case previously derived [4], i.e., when there was no influence of a piston, is given in Appendix D. This function, called $F_0(\sigma, m_1)$, is obtained from Eq. (3.4.4) when $K = K(\sigma, m_1, 0)$ as obtained from Eq. (2.3.5).

The tables may be used in several ways, e.g., suppose there is a section of variable area connecting two ducts of constant but unequal area, say

A_1 and A_2, then for given m_1,

$$\frac{F(\sigma_2, m_1)}{F(\sigma_1, m_1)} = \frac{A_2}{A_1} .$$

(3.4.5)

If A_1, A_2 and σ_1 are known, σ_2 may be obtained by interpolating inversely from the tables.

After passage through the transition section, the shock wave travels with this altered shock strength, σ_2, and ultimately becomes uniform. Conversely, given σ_1, σ_2, m_1 and A_1, the tables may be utilized to determine what A_2 would give the desired σ_2. Eq. (3.4.5) could also be considered as one for m_1 with σ_1, σ_2, A_1 and A_2 prescribed.

The tables for $F_0(\sigma, m_1)$ are valid for arbitrary area variations, so that converging cylindrical and spherical shocks may be treated by choosing channels with areas proportional to D and D^2, respectively, where D is the distance from the axis or point of symmetry.

Finally, from the tables of Appendix E, which are valid for areas proportional to D and D^2 only, it is seen that the asymptotic limits of K_c and K_s are independent of the applied field, so that there is the result

Near the axis or point of symmetry, the strengths of converging cylindrical and spherical piston-driven magnetohydrodynamic shock waves are independent of the applied field and are given by conventional gas dynamic theory.

Chapter 4

Flows with Heat Addition

4.1 The Linearized Equations

Another problem which may be treated by the present method is to determine the effects of small heat addition and area variations on the otherwise uniform flow of a perfectly conducting fluid [32]. The heat addition produces entropy variations and, since it is generally the rate of change of entropy along a trajectory that is known, it is convenient to use (u, P, B, s) as dependent variables [33]. With this choice of dependent variables, only material derivatives of the entropy appear, and the governing equations are

$$P_t + u P_x + \gamma P u_x - P\left[\frac{f^*}{c_v} - \frac{\gamma u A_x}{A}\right] = 0$$

(4.1.1)

$$\varrho u_t + \varrho u u_x + P_x + \frac{B B_x}{\mu} = 0$$

(4.1.2)

$$B_t + u B_x + B u_x = 0 ,$$

(4.1.3)

$$s_t + u s_x = f^*$$

(4.1.4)

where f^* is the rate of entropy production (due to the heat addition).

The characteristics of the system of Eq. (4.1.1)—(4.1.4) are, of course, given by

$$\frac{dx}{dt} = u, \quad u + \omega, \quad u - \omega$$

and the system may be written in the following characteristic form

$$P_\beta + \varrho\omega u_\beta + \varrho(\omega^2 - c^2)\frac{B_\beta}{B} - P\left[\frac{f^*}{c_v} - \frac{\gamma u A_x}{A}\right] = 0 \qquad (4.1.5)$$

$$P_\alpha - \varrho\omega u_\alpha + \varrho(\omega^2 - c^2)\frac{B_\alpha}{B} - P\left[\frac{f^*}{c_v} - \frac{\gamma u A_x}{A}\right] = 0 \qquad (4.1.6)$$

$$\frac{P_\xi}{\gamma P} - \frac{B_\xi}{B} - \left[\frac{f^*}{\gamma c_v} - \frac{u A_x}{A}\right] = 0 \qquad (4.1.7)$$

$$S_\xi = f^* . \qquad (4.1.8)$$

Approximate solutions to the characteristic system could be obtained by finite-difference techniques.

To this point, the discussion has been quite general, and the equations apply to an arbitrary quasi-one-dimensional flow with heat addition. But, if it is assumed now that the cross-sectional area variations and heat addition are small and may be considered as small perturbations in an otherwise uniform base flow, it follows from Eq. (4.1.7) that B/ϱ is constant along each particle path and for an initially uniform flow, B/ϱ is constant throughout the flow. Then, Eqs. (4.1.5)—(4.1.6) lead to

$$\frac{u}{2} + \frac{1}{2\gamma}\int\frac{\omega d P}{P} = \alpha \qquad (4.1.9)$$

$$-\frac{u}{2} + \frac{1}{2\gamma}\int\frac{\omega d P}{P} = \beta \qquad (4.1.10)$$

which is merely a different form of the generalized Riemann invariants as defined by Eqs. (1.2.16)—(1.2.17). Of course, Eqs. (4.1.9)—(4.1.10) could have been obtained directly from Eqs. (1.2.16)—(1.2.17).

Consequently, under exactly the same assumption as was made in the previous chapters, i.e., the constancy of B/ϱ throughout the flow, the system of Eqs. (4.1.1)—(4.1.4) is equivalent to the following system

$$\frac{\omega}{\gamma P}\left[P_t + (u + \omega)P_x\right] + \left[u_t + (u + \omega)u_x\right] - \\ -\frac{\omega}{\gamma}\left[\frac{f^*}{c_v} - \frac{\gamma u A_x}{A}\right] = 0 \qquad (4.1.11)$$

$$\frac{\omega}{\gamma P}\left[P_t + (u - \omega)P_x\right] - \left[u_t + (u - \omega)u_x\right] - \\ -\frac{\omega}{\gamma}\left[\frac{f^*}{c_v} - \frac{\gamma u A_x}{A}\right] = 0 \qquad (4.1.12)$$

$$S_t + u S_x = f^* . \qquad (4.1.4)$$

When Eqs. (4.1.11), (4.1.12) and (4.1.4) are linearized in the neighborhood of a known constant isentropic solution, denoted by the subscript zero, the following system of linear equations for the terms of first order, denoted by the subscript one, is obtained

$$R_t + (u_0 + \omega_0) R_x = \frac{\omega_0}{2\gamma} \left[\frac{f_1^*}{c_v} - \frac{\gamma u_0 A_1'(x)}{A_0} \right] \tag{4.1.13}$$

$$S_t + (u_0 - \omega_0) S_x = \frac{\omega_0}{2\gamma} \left[\frac{f_1^*}{c_v} - \frac{\gamma u_0 A_1'(x)}{A_0} \right] \tag{4.1.14}$$

$$s_{1t} + u_0 s_{1x} = f_1^* \tag{4.1.15}$$

where

$$R_1 = \frac{u_1}{2} + \frac{\omega_0 P_1}{2\gamma P_0}$$

$$S = -\frac{u_1}{2} + \frac{\omega_0 P_1}{2\gamma P_0}$$

are the first order generalized Riemann invariants written in terms of u_1 and P_1. For simplicity, let $f_1^* = f_1^*(t)$ alone; then, exactly as in Chapter 1, the general solution of Eqs. (4.1.13)–(4.1.15) may be written as

$$s_1 = \Gamma(\Psi_0) + \int f_1^*(t)\, dt \tag{4.1.16}$$

$$R = F\left[x - (u_0 + \omega_0)t\right] + \frac{\omega_0}{2\gamma c_v} \int f_1^*(t)\, dt - \frac{u_0 \omega_0 A_1(x)}{2 A_0 (u_0 + \omega_0)} \tag{4.1.17}$$

$$S = G\left[x - (u_0 - \omega_0)t\right] + \frac{\omega_0}{2\gamma c_v} \int f_1^*(t)\, dt - \frac{u_0 \omega_0 A_1(x)}{2 A_0 (u_0 - \omega_0)} \tag{4.1.18}$$

with F, G and Γ arbitrary functions. For the case of no heat addition, these solutions are equivalent to Eqs. (1.2.30)–(1.2.32), and, in the limit of vanishing magnetic field, these solutions reduce to solutions of the corresponding problem in conventional gas dynamics. An illustration of this latter fact will be given in the next section.

4.2 Extension of Stocker's Work

In many technological applications, it is of interest to determine the modifying effects due to the addition of heat to a gas flow. When heat is added to a gas flow, entropy changes are produced and, if the flow is subsonic (supersonic), it accelerates (decelerates). Because the solution in the subsonic case predicts some rather anomalous behavior, the solution must be interpreted with great care. For example, suppose heat is continuously added to a steady, subsonic one-dimensional gas flow in a uniform duct in such a way that the flow remains one-dimensional; then, the following results were stated by several investigators: (i) although the flow speed accelerates, a subsonic flow cannot become supersonic but rather always approaches the sonic state; (ii) the temperature T increases monotonically to a maximum for $M = 1/\gamma^{\frac{1}{2}}$, where M is the Mach number, but then starts to decrease, i.e., the temperature actually decreases when heat is added if $1 > M > 1/\gamma^{\frac{1}{2}}$. A rather heated controversy was generated by the many different hypotheses made in attempting to justify these results. The correct interpretation depends, however, on a careful examination of the transient effects introduced in the initially steady flow by the heat addition and, in particular, on the effects of the transients on the boundary conditions at the entrance and exit of the region of heating. This was carried out by FOA and RUDINGER [34], [35] who showed that heat addition: (i) modified the flow parameters upstream and downstream of the heating region; (ii) M may increase or decrease depending on the subsidiary conditions; (iii) the static temperature increases monotonically with increasing M for $1 > M > 1/\gamma^{\frac{1}{2}}$. Their discussion was based on an initially steady flow in a duct finite in length, and it was possible to determine the final modified state without considering the intermediary flow.

A more direct approach was presented in an interesting paper by STOCKER [13], who used a small perturbation analysis to determine the effects of transients introduced by adding heat to an initially uniform subsonic flow in an infinite duct of uniform cross-sectional area. Further, it was shown that this solution could be used to discuss the case of a finite duct where the solution depends on the subsidiary conditions at the ends of the duct.

In the present section, it is shown that STOCKER's discussion may be extended to magnetohydrodynamic flows subjected to a transverse magnetic field, and further, in such a form that in the limit of vanishing magnetic field, the solution in the magnetic case reduces precisely to the solution in the non-magnetic case [37]. Specifically, the transients due to uniform heat addition along a finite section of an infinite duct of uniform cross-sectional area are determined by the linearized analysis of section 4.1 which is based on small heat addition. Throughout, it is assumed that the initially uniform flow is sub-true-sonic. Although the solution obtained ceases to be valid in the neighborhood of true-sonic flow, and this difficulty is due to the inadequacies of the linearization which approximates the characteristics by two rectilinear and parallel families of lines, a solution valid for *all* flow speeds may be obtained by making a more detailed examination of the negative characteristics.

For small perturbations, denoted by the subscript one, of an initially uniform flow, denoted by the subscript zero, in a uniform duct, the pertinent equations are Eqs. (4.1.17)—(4.1.18) which show that

$$R = \frac{u_1}{2} + \frac{\omega_0 P_1}{2\gamma P_0} = \frac{\omega_0}{2\gamma c_v} \int f_1^*(t)\, dt + \text{constant} \qquad (4.2.1)$$

on advancing characteristics

$$x - (u_0 + \omega_0)t = \text{constant}.$$

$$S = -\frac{u_1}{2} + \frac{\omega_0 P_1}{2\gamma P_0} = \frac{\omega_0}{2\gamma c_v} \int f_1^*(t)\, dt + \text{constant} \qquad (4.2.2)$$

on receding characteristics

$$x - (u_0 - \omega_0)t = \text{constant} .$$

In regions where there is no heating, Eqs. (4.2.1)—(4.2.2) are valid with $f_1^*(t) = 0$.

The sound speed perturbation may be determined from

$$\frac{c}{c_0} = \left(\frac{P}{P_0}\right)^{(\gamma-1)/2\gamma} \exp\left[(s - s_0)/2\gamma c_v\right]$$

which gives

$$\frac{c_1}{c_0} = \frac{(\gamma - 1) P_1}{2\gamma P_0} + \frac{s_1}{2\gamma c_v} .$$

The solution for the first order generalized Riemann invariants, Eqs. (4.2.1)—(4.2.2) may be used to solve the problem of uniform heat addition to the initially uniform flow of a gas in an infinite duct. Specifically, it is assumed that at $t = 0$, a gas is flowing through a uniform duct infinitely long in both directions, but for $t \geqq 0$, heat is added uniformly over the section extending from $x = 0$ to $x = \xi_1$ and that the flow remains one-dimensional. Further, it is assumed that

$$\frac{\omega_0 f_1^*(t)}{\gamma c_v} = \lambda_1 ,$$

a constant. This follows because in the linearized approximation, a uniform rate of entropy production is equivalent to a uniform rate of heat addition. Consequently, from Eqs. (4.2.1)—(4.2.2),

$$R - \frac{\lambda_1 t}{2} = \text{constant on advancing characteristics} , \qquad (4.2.3)$$

$$S - \frac{\lambda_1 t}{2} = \text{constant on receding characteristics} . \qquad (4.2.4)$$

This is the direct extension to the magnetic case of the problem considered by STOCKER, and its solution is obtained in a similar manner. The (x, t) plane is divided into regions as shown in Fig. 7, which is taken from STOCKER's paper [13]. The region contained between the lines $x = 0$ and $x = \xi_1$ is the heating region and Eqs. (4.2.3)—(4.2.4) apply therein. Outside this region $R(S)$ is constant on an advancing (receding) characteristic. The lines OD and LC are the advancing characteristics

$$x - (u + \omega)t = 0 \quad \text{and} \quad x - (u + \omega)t = \xi_1 .$$

The lines OA and LB are the receding characteristics

$$x - (u - \omega)t = 0 \quad \text{and} \quad x - (u - \omega)t = \xi_1 .$$

The lines OM and LN are the particle paths

$$x - ut = 0 \quad \text{and} \quad x - ut = \xi_1 \,.$$

For simplicity the subscript zero has been dropped from the base flow.

The values of R and S throughout the (x, t) plane may be found by the use of Eqs. (4.2.3)—(4.2.4). Since $R(S)$ is constant along an advancing (receding) characteristic and $R = 0$, $S = 0$ in regions 1 and 15, the values

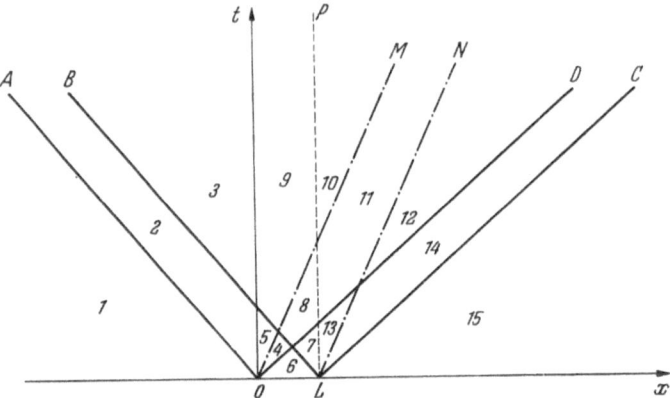

Fig. 7. The (x, t) diagram for uniform heat addition to an initially uniform flow. Heat is added over the section extending from O to L. The lines OD and LC are advancing characteristics. The lines OA and LB are receding characteristics. The lines OM and LN are particle paths

of R and S in the other regions may be found by considering characteristics which originate therein. This gives:

In regions 10, 11, 12, 13, 14

$$S = 0 \,.$$

In regions 7, 8, 9

$$S = \frac{\lambda_1 (\xi_1 - x)}{2 (\omega - u)} \,.$$

In regions 4, 5, 6

$$S = \frac{\lambda_1 t}{2} \,.$$

In region 2

$$S = \frac{\lambda_1}{2} \left[t + \frac{x}{\omega - u} \right] \,.$$

In region 3

$$S = \frac{\lambda \xi_1}{2 (\omega - u)} \,.$$

In regions 1, 2, 3

$$R = 0 \,.$$

In regions 4, 5, 8, 9

$$R = \frac{\lambda_1 x}{2(u + \omega)} \; .$$

In regions 6, 7

$$R = \frac{\lambda_1 t}{2} \; .$$

In regions 13, 14

$$R = \frac{\lambda_1}{2}\left[t + \frac{\xi_1 - x}{u + \omega}\right].$$

In regions 10, 11, 12

$$R = \frac{\lambda_1 \xi_1}{2(u + \omega)} \; .$$

These relations allow the determination of u_1 and P_1 in the regions of interest. To determine c_1, ω_1, T_1, M_1 and N_1, it is necessary to know s_1. In regions 1, 2, 3, 12, 14 and 15 and on the line OL, $s_1 = 0$. Starting from a region wherein $s_1 = 0$ and integrating Eq. (4.1.15) along particle paths, the perturbed entropy distribution may be determined, and this gives:

In regions 4, 6, 7 and 8

$$\frac{s_1}{c_v} = \frac{\gamma \lambda_1 t}{\omega} \; .$$

In regions 5 and 9

$$\frac{s_1}{c_v} = \frac{\gamma \lambda_1 x}{u \omega} \; .$$

In region 10

$$\frac{s_1}{c_v} = \frac{\gamma \lambda_1 \xi_1}{u \omega} \; .$$

In regions 11 and 13

$$\frac{s_1}{c_v} = \frac{\gamma \lambda_1 (\xi_1 - x + ut)}{u \omega} \; .$$

Consequently, the perturbed quantities in the various regions are:

In region 1

$$u_1 = 0, \quad P_1 = 0, \quad c_1 = 0, \quad M_1 = 0 .$$

In region 2

$$u_1 = -\frac{\lambda_1}{2}\left[t + \frac{x}{\omega - u}\right],$$

$$\frac{P_1}{P} = \frac{\gamma \lambda_1}{2\omega}\left[t + \frac{x}{\omega - u}\right],$$

$$\frac{c_1}{c} = \frac{(\gamma - 1)\lambda_1}{4\omega}\left[t + \frac{x}{\omega - u}\right],$$

$$M_1 = -\lambda_1\left[\frac{2\omega + (\gamma - 1)u}{4\omega c}\right]\left[t + \frac{x}{\omega - u}\right].$$

In region 3

$$u_1 = -\frac{\lambda_1 \xi_1}{2(\omega - u)},$$

$$\frac{c_1}{c} = \frac{(\gamma - 1)\lambda_1 \xi_1}{4\omega(\omega - u)},$$

$$\frac{P_1}{P} = \frac{\gamma \lambda_1 \xi_1}{2\omega(\omega - u)},$$

$$M_1 = -\frac{\lambda_1 \xi_1 [2\omega c + (\gamma - 1)u]}{4\omega c(\omega - u)}.$$

In region 4

$$u_1 = \frac{\lambda_1}{2}\left[\frac{x}{u + \omega} - t\right],$$

$$\frac{P_1}{P} = \frac{\xi \lambda_1}{2\omega}\left[t + \frac{x}{u + \omega}\right],$$

$$\frac{c_1}{c} = \frac{\lambda_1}{4\omega}\left[(\gamma + 1)t + (\gamma - 1)\frac{x}{u + \omega}\right],$$

$$M_1 = -\frac{\lambda_1}{4\omega c}\left\{[2\omega + (\gamma + 1)u]t + \frac{x[(\gamma - 1)u - 2\omega]}{u + \omega}\right\}.$$

In region 5

$$u_1 = \frac{\lambda_1}{2}\left[\frac{x}{u + \omega} - t\right],$$

$$\frac{P_1}{P} = \frac{\gamma \lambda_1}{2\omega}\left[\frac{x}{u + \omega} + t\right],$$

$$\frac{c_1}{c} = \frac{(\gamma - 1)\lambda_1 t}{4\omega} + \frac{\lambda_1 x[2\omega + (\gamma + 1)u]}{4u\omega(u + \omega)},$$

$$M_1 = -\frac{\lambda_1}{4\omega c}\left\{[2\omega + (\gamma - 1)u]t + \frac{xu(\gamma + 1)}{u + \omega}\right\}.$$

In region 6

$$u_1 = 0,$$

$$\frac{P_1}{P} = \frac{\gamma \lambda_1 t}{\omega},$$

$$\frac{c_1}{c} = \frac{\gamma \lambda_1 t}{2\omega},$$

$$M_1 = -\frac{\gamma \lambda_1 u t}{2\omega c}.$$

In region 7

$$u_1 = \frac{\lambda_1}{2}\left[t + \frac{x - \xi_1}{\omega - u}\right],$$

$$\frac{c_1}{c} = \frac{\lambda_1}{4\omega}\left[(\gamma + 1)t + \frac{(\gamma - 1)(\xi_1 - x)}{\omega - u}\right],$$

$$\frac{P_1}{P} = \frac{\gamma \lambda_1}{2\omega}\left[t + \frac{\xi_1 - x}{\omega - u}\right],$$

$$M_1 = \frac{\lambda_1}{4\omega c}\left[t\{2\omega - (\gamma + 1)u\} - \{2\omega + (\gamma - 1)u\}\frac{(\xi_1 - x)}{\omega - u}\right].$$

In regions 8 and 9

$$u_1 = \frac{\lambda_1 [2\omega x - (u+\omega)\xi_1]}{2(\omega^2 - u^2)},$$

$$\frac{P_1}{P} = \frac{\gamma \lambda_1 [(u+\omega)\xi_1 - 2ux]}{2\omega(\omega^2 - u^2)}.$$

In region 8

$$\frac{c_1}{c} = \frac{\lambda_1 t}{2\omega} + \frac{(\gamma - 1)\xi_1 \lambda_1}{4\omega(\omega - u)} - \frac{(\gamma - 1)\lambda_1 ux}{2\omega(\omega^2 - u^2)},$$

$$M_1 = -\frac{u\lambda_1 t}{2\omega c} + \frac{\lambda_1 x [2\omega^2 + (\gamma - 1)u^2]}{2\omega c(\omega^2 - u^2)} - \frac{\lambda_1 \xi_1 [2\omega + (\gamma - 1)u]}{4\omega c(\omega - u)}.$$

In region 9

$$\frac{c_1}{c} = \frac{(\gamma - 1)\lambda_1 \xi_1}{4\omega(\omega - u)} + \frac{(\omega^2 - \gamma u^2)\lambda_1 x}{2\omega u(\omega^2 - u^2)},$$

$$M_1 = \frac{\lambda_1 x(\omega^2 + \gamma u^2)}{2\omega c(\omega^2 - u^2)} - \frac{\lambda_1 \xi_1 [2\omega + (\gamma - 1)u)]}{4\omega c(\omega - u)}.$$

In regions 10, 11, 12

$$u_1 = \frac{\lambda_1 \xi_1}{2(u+\omega)},$$

$$\frac{P_1}{P} = \frac{\gamma \lambda_1 \xi_1}{2\omega(u+\omega)}.$$

In region 10

$$\frac{c_1}{c} = \frac{[(\gamma + 1)u + 2\omega]\lambda_1 \xi_1}{4\omega u(u+\omega)},$$

$$M_1 = -\frac{(\gamma + 1)u\lambda_1 \xi_1}{4\omega(\omega + u)}.$$

In region 11

$$\frac{c_1}{c} = \frac{\lambda_1 \xi_1 [(\gamma + 1)u + 2\omega]}{4u\omega(u+\omega)} - \frac{\lambda_1 x}{2u\omega} + \frac{\lambda_1 t}{2\omega}.$$

$$M_1 = -\frac{(\gamma + 1)u\lambda_1 \xi_1}{4\omega c(u+\omega)} + \frac{\lambda_1 x}{2\omega c} - \frac{\lambda_1 ut}{2\omega c}.$$

In region 12

$$\frac{c_1}{c} = \frac{(\gamma - 1)\lambda_1 \xi_1}{4\omega(u+\omega)},$$

$$M_1 = \frac{\lambda_1 \xi_1 [2\omega - (\gamma - 1)u]}{4\omega c(u+\omega)}.$$

In regions 13 and 14

$$u_1 = \frac{\lambda_1}{2}\left[t + \frac{\xi_1 - x}{u+\omega}\right],$$

$$\frac{P_1}{P} = \frac{\gamma \lambda_1}{2\omega}\left[t + \frac{\xi_1 - x}{u+\omega}\right].$$

In region 13

$$\frac{c_1}{c} = \frac{(\gamma + 1)\lambda_1 t}{4\omega} + \frac{\lambda_1(\xi_1 - x)\,[(\gamma + 1)u + 2\omega]}{4\omega u(u + \omega)},$$

$$M_1 = \lambda_1 t \left[\frac{2\omega - (\gamma + 1)u}{4\omega c}\right] - \lambda_1(\xi_1 - x)\left[\frac{(\gamma + 1)u}{4\omega c(u + \omega)}\right].$$

In region 14

$$\frac{c_1}{c} = \frac{(\gamma - 1)\lambda_1}{4\omega}\left[t + \frac{\xi_1 - x}{u + \omega}\right],$$

$$M_1 = \frac{\lambda_1[2\omega - (\gamma - 1)u]}{4\omega c}\left[t + \frac{\xi_1 - x}{u + \omega}\right].$$

In region 15

$$u_1 = 0,$$

$$P_1 = 0,$$

$$c_1 = 0,$$

$$M_1 = 0.$$

The values of N_1, ω_1, T_1 in any of the above regions may be obtained from the relations

$$\frac{\omega_1}{\omega} = \left[\frac{q^2 + \gamma - 2}{(\gamma - 1)q^2}\right]\frac{c_1}{c},$$

$$qN_1 = M_1 + \frac{M(\gamma - 2)(q^2 - 1)c_1}{(\gamma - 1)q^2\,c},$$

$$\frac{T_1}{T} = \frac{2c_1}{c}$$

where $N = u/\omega$.

In the limit of vanishing magnetic field, these results reduce precisely to those obtained by STOCKER.

As STOCKER points out, the unsteady effects are confined to the regions: (i) 2, which represents a pressure wave propagating upstream; (ii) 11, which represents an interface propagating downstream; (iii) 14, which represents a pressure wave propagating downstream. Consequently, the flow resulting from the heat addition never becomes steady, though the flow in any finite section asymptotically becomes steady and, in a sense, the flow in regions 3, 9 and 10 is the final steady flow resulting from the heating of the initially uniform flow.

In regions 3, 9 and 10, $c_1 > 0$, $P_1 > 0$, $M_1 < 0$ and $T_1 > 0$, so that an addition of heat cannot result in a drop in temperature, and an increase in total heat added to the fluid passing through the heating section cannot cause a fall in temperature at the exit from this section. Within the framework of the linearized analysis, the final states in regions 3 and 9 depend only on the product $\lambda_1 \xi_1$, i.e., on the total heat added but not directly on the length over which the addition takes place.

In the heating region 9, $\frac{\partial T}{\partial x} < 0$ if $M > q/\gamma^{\frac{1}{2}}$ or, equivalently, if $N > \gamma^{-\frac{1}{2}}$. This is because the gas flowing through the heating section undergoes a pressure drop and the heat loss from the concomitant expansion may exceed the gain from the external source. In the unsteady flow regions 7, 8, 10 and 11, $\frac{\partial T}{\partial x} < 0$, irrespective of the Mach number.

The solution of this section may be used to discuss the case of a duct of finite length. The solution is identical to that previously obtained until the advancing or receding waves encounter an end of the duct where reflection takes place. Again, the solution in the magnetic case follows directly from STOCKER's solution for the non-magnetic case and will not be included.

4.3 Modified Perturbation Theory

The linearized analysis of section 4.2 led to a solution which was singular for true-sonic speed, i.e., for $u = \omega$, in the neighborhood of which the disturbance S tends to accumulate and does not propagate through the channel. Exactly as in Chapter 2, this difficulty is due to the linearization, which approximates the negative characteristics $dx/dt = u - \omega$ by the rectilinear family $x - (u - \omega)t = $ constant, and again a more detailed examination of the receding characteristics must be made.

From Eqs. (4.2.1)−(4.2.2), it is seen that the values of u_1 and P_1 depend on the time taken by a small disturbance to pass through the heating region. This is true because the function f_1^* was assumed constant. Although the time will always be small for the advancing characteristics, it becomes large for the receding characteristics when the base flow is near true-sonic. Thus, it may be assumed that the advancing characteristics and particle paths are each rectilinear and parallel, viz.,

$$x - (u + \omega)t = \text{constant} ,$$

$$x - ut = \text{constant} ,$$

while the negative characteristics are given by

$$dx/dt = u - \omega + u_1 - \omega_1 . \tag{4.3.1}$$

Let the solution of Eq. (4.3.1) be written as

$$\eta(x, t) = \text{constant} .$$

In a heating region, $R - \lambda_1 t/2$ is constant on an advancing characteristic and thus a function of $x - (u + \omega)t$; $S - \lambda_1 t/2$ is constant on a receding characteristic and thus a function of η.

The explicit solution for the fifteen regions in the (x, t) plane will not be given; rather, only the regions 7, 13 and 14 will be considered. These are the cases considered by STOCKER.

In region 14, it is known that

$$R = \frac{\lambda_1}{2}\left[t + \frac{\xi_1 - x}{u + \omega}\right], \quad S = 0, \quad s_1 = 0.$$

Thus, the receding characteristics are determined from

$$\left(\frac{\partial x}{\partial t}\right)_\eta = u - \omega + u_1 - \omega_1 = u - \omega + \frac{\lambda_1}{4}\left[\frac{q^2 - \gamma + 2}{q^2}\right]\left[t + \frac{\xi_1 - x}{u + \omega}\right]. \quad (4.3.2)$$

It is convenient to choose η so that it is approximately equal to $x - (u - \omega)t$, i.e., choose

$$\eta = x - (u - \omega)t + \lambda_1 \eta_1(x, t) + 0(\lambda_1^2). \quad (4.3.3)$$

Substituting Eq. (4.3.3) into Eq. (4.3.2) and carrying out the integration, there results

$$x = (u - \omega)t + \frac{\lambda_1[q^2 - \gamma + 2]}{4q^2}\left\{\frac{\omega t^2}{u + \omega} + \frac{t(\xi_1 - \eta)}{u + \omega}\right\} + h(\eta) + 0(\lambda_1^2 t).$$

The function of integration $h(\eta)$ is not completely arbitrary since by the definition of η, it follows that

$$h(\eta) = \eta + \lambda_1 h_1(\eta) + 0(\lambda_1^2)$$

where $h_1(\eta)$ is arbitrary. It may be determined by requiring η to be continuous. In region 15, the receding characteristics are rectilinear and $\eta = x - (u - \omega)t$. When η is required to be continuous across LC, i.e., $x = (u + \omega)t + \xi_1$, it is found that

$$h_1(\eta) = \frac{[q^2 - \gamma + 2](\eta - \xi_1)^2}{16q^2 \omega(u + \omega)}.$$

Thus

$$x = (u - \omega)t + \frac{\lambda_1(q^2 - \gamma + 2)t[\omega t + \xi_1 - \eta]}{4q^2(u + \omega)} + \eta + \frac{\lambda_1[q^2 - \gamma + 2](\eta - \xi_1)^2}{16q^2 \omega(u + \omega)}.$$

are the negative characteristics in region 14.

In region 13,

$$R = \frac{\lambda_1}{2}\left[t + \frac{\xi_1 - x}{u + \omega}\right], \quad S = 0,$$

$$\frac{s_1}{\gamma c_v} = \frac{\lambda_1(\xi_1 - x + ut)}{u\omega}.$$

Thus,

$$\left(\frac{\partial x}{\partial t}\right)_\eta = u - \omega + \frac{\lambda_1(q^2 + 2 - \gamma)}{4q^2}\left[t + \frac{\xi_1 - x}{u + \omega}\right] - \frac{\lambda_1(q^2 + \gamma - 2)(\xi_1 - x + ut)}{2(\gamma - 1)q^2 u}$$

so that by the same procedure

$$x = (u - \omega)t + \frac{\lambda_1(q^2 - \gamma + 2)\,t\,(\omega t + \xi_1 - \eta)}{4q^2(u + \omega)} -$$

$$- \frac{\lambda_1(q^2 + \gamma - 2)t}{2u(\gamma - 1)q^2}\left[\frac{\omega t}{2} + \xi_1 - \eta\right] + m(\eta),$$

where $m(\eta) = \eta + \lambda_1 m_1(\eta) + O(\lambda_1^2)$,

where $m_1(\eta)$ is arbitrary.

By assuming that η is continuous across the line LN, i.e., $x = ut + \xi_1$, it is found that

$$m_1(\eta) = h_1(\eta) - \frac{(q^2 + \gamma - 2)\,(\eta - \xi_1)^2}{4(\gamma - 1)q^2 u\omega},$$

i.e.,

$$m_1(\eta) = \frac{(\eta - \xi_1)^2}{4\omega}\left\{\frac{q^2 - \gamma + 2}{4q^2(u + \omega)} - \frac{(q^2 + \gamma - 2)}{(\gamma - 1)q^2 u}\right\}.$$

In region **7**, the modification of the characteristics leads also to different values of S. It is known that

$$R = \frac{\lambda_1 t}{2},\quad \frac{S_1}{2\gamma c_v} = \frac{\lambda_1 t}{2\omega}.$$

It is known that $S - \lambda_1 t/2$ is constant along $\eta = $ constant. Thus, let

$$S = \frac{\lambda_1[t - \tau(\eta)]}{2}$$

where $\tau(\eta)$ is value of t where the characteristic meets the line $x = \xi_1$. Thus

$$S = \frac{\lambda_1}{2}\left[t + \frac{\eta - \xi_1}{u - \omega}\right],$$

so that

$$u_1 = -\frac{\lambda_1(\eta - \xi_1)}{2(u - \omega)},$$

$$\frac{\omega_1}{\omega} = \frac{\gamma\lambda_1 t[q^2 + \gamma - 2]}{2\omega(\gamma - 1)q^2} + \frac{[q^2 + \gamma - 2]\lambda_1(\eta - \xi_1)}{4\omega(u - \omega)q^2}.$$

Thus

$$\left(\frac{\partial x}{\partial t}\right)_\eta = u - \omega - \frac{\lambda_1(\eta - \xi_1)\,[3q^2 + \gamma - 2]}{4q^2(u - \omega)} - \frac{[q^2 + \gamma - 2]\gamma\lambda_1 t}{2(\gamma - 1)q^2}$$

and

$$x = (u - \omega)t - \frac{\lambda_1(\eta - \xi_1)\,(3q^2 + \gamma - 2)t}{4q^2(u - \omega)} -$$

$$- \frac{(q^2 + \gamma - 2)\gamma\lambda_1 t^2}{4(\gamma - 1)q^2} + \eta + \lambda_1 n_1(\eta) + O(\lambda_1^2).$$

The arbitrary function n_1 is determined by choosing η to be continuous across the line $x = \xi_1$. This gives

$$n_1(\eta) = m_1(\eta) + \frac{(\xi_1 - \eta)^2}{4q^2(u - \omega)^2}\left\{\frac{q^2(3 - 2\gamma) + \gamma - 2}{\gamma - 1} + \right.$$

$$+ \left. \frac{u^2[(\gamma - 3)q^2 + 2 + \gamma - \gamma^2] - u\omega(q^2 + \gamma - 2) + \omega^2(q^2 + \gamma - 2)}{(\gamma - 1)u(u + \omega)}\right\}.$$

STOCKER has given a comparison of the solutions in region 9 obtained through the use of the modified perturbation theory and by nonlinear steady-state theory as obtained by numerical integration of the equations given in [36] under the assumptions that the upstream (downstream) boundary condition is $R = 0$ $(S = 0)$. In all cases, the modified linear solution was a reasonable approximation to the nonlinear solution, even in the cases where the linear solution failed. The same conclusions may be expected to obtain in the present case.

Chapter 5

Simple Wave Flows
5.1 Introductory Comments

In isentropic flows, simple waves are defined by the condition that the fluid velocity, the local speed of sound and the magnetic field are constant along any curve of one family of characteristics. As a consequence, this family of characteristics is rectilinear. For non-isentropic flows, the same condition cannot be assumed for the entropy. This is due to the fact that the entropy is constant along a streamline. In addition, if the entropy is constant along a family of characteristics, which is distinct from the streamlines, then the flow is isentropic. It might be assumed that a possible solution would be to assume that the entropy is not constant along a simple wave, but that the fluid velocity, local sound speed and magnetic field are constant along a simple wave. But from the characteristic form of the basic equations, viz., Eqs. (1.2.4)—(1.2.10), it is seen that such an assumption is not possible, so that simple waves (in the above sense) do not exist in non-isentropic flow. However, non-isentropic perturbations of simple waves have been shown to exist [30].

For conventional one-dimensional gas dynamic flows, the non-isentropic perturbation of a centered simple wave and the isentropic perturbation of an arbitrary (non-centered) simple wave were determined by GERMAIN and GUNDERSEN [2], and a further discussion given by GUNDERSEN [5], [6]. The non-isentropic perturbation of an arbitrary simple wave was determined by GUNDERSEN [38], and the results of these papers were used to discuss centered and arbitrary simple wave flow in ducts with small non-uniform cross-sectional area perturbations, viz., linear and quadratic area distributions [39].

ROSCISZEWSKI [41] has used the Germain-Gundersen solution for the isentropic perturbation of a centered simple wave to consider centered simple wave flow in a duct with a small linear area variation; however, there is no reference to that effect, nor, for that matter, is there any reference to the earlier and more extensive discussion of effectively the same problem by GUNDERSEN [39]. FRIEDMAN [42] has used a minor variation of the results of [2], [6] as a basis for a discussion of cylindrical and spherical shock waves. References [6], [41] and [42] all appeared in the same journal.

Recently, it was shown that the gas dynamic theory [2] could be extended to magnetohydrodynamic flows, and the non-isentropic perturbation of a centered magnetohydrodynamic simple wave was determined [30]. Hydromagnetic simple wave flow in non-uniform ducts was discussed, and the non-isentropic perturbation of a non-centered simple wave determined [40]. These solutions [30], [40] were limited to a monatomic fluid, and it is the primary purpose of this chapter to extend these results to an arbitrary value of the adiabatic index. Since simple wave flow in non-uniform ducts has been discussed previously [40], the present discussion will be limited to determining the non-isentropic perturbation of a centered and non-centered simple wave in a uniform duct. The effects of cross-sectional area perturbations may be included by adding particular solutions, obtainable as quadratures, of the complete perturbation equations.

In Chapter 1, it was shown that in the limit of vanishing magnetic field, the non-isentropic perturbation of an initially uniform flow reduced precisely to the corresponding problem in conventional gas dynamics. The same situation *does not* obtain for an initially simple wave flow. However, the *isentropic* perturbation of a simple wave *does* reduce exactly to the corresponding problem in conventional gas dynamics in the limit of vanishing magnetic field.

For a monatomic gas, it is even possible to give an *exact* solution for centered simple wave flow [40], and this is discussed in section 5.2. Further, the interaction of simple waves may be reduced to the solution of a second order linear partial differential equation somewhat similar to the Euler-Poisson Equation [43], and this is discussed in section 5.2. With the exception of this section, the results of the remainder of the chapter are valid for arbitrary values of the adiabatic index.

5.2 The Monatomic Fluid

It is assumed that the wave is produced by the impulsive withdrawal of a piston in a tube filled with a monatomic gas initially at rest. Let the origin of the (x, t) coordinate system be taken at the initial position of the piston, so that the resultant (forward-facing) rarefaction wave is centered at the origin and characterized by $\beta = \beta_0$. On each characteristic, $dx/dt = u + \omega$, (u, c, B) are constant, and these characteristics are straight lines in the (x, t) plane. The wave may be represented by

$$x = (u + \omega)t, \tag{5.2.1}$$

$$-\frac{u}{2} + \frac{(1 + kc)^{3/2}}{k} = \beta_0 = \frac{(1 + kc_r)^{3/2}}{k} \tag{5.2.2}$$

where the subscript r denotes stagnation quantities. The wave propagates with speed ω_r into the gas at rest.

From Eqs. (5.2.1)—(5.2.2), it is possible to *explicitly* determine the flow parameters in the wave. The substitution $d = 1 + kc$ leads to the cubic equation

$$d^3 - \frac{d}{3} - \frac{2k}{3}\left(\beta_0 + \frac{x}{2t}\right) = 0$$

which has the solution

$$d = a_1^{1/3} + a_2^{1/3}$$

where

$$a_1 = a_3 + \left(a_3^2 - \frac{1}{729}\right)^{\frac{1}{2}},$$

$$a_2 = a_3 - \left(a_3^2 - \frac{1}{729}\right)^{\frac{1}{2}},$$

$$a_3 = \frac{k}{3}\left(\beta_0 + \frac{x}{2t}\right).$$

From this result and the definition of the characteristic parameters, it follows that

$$kc = a_1^{2/3} + a_2^{2/3} - \frac{7}{9} = \left[\frac{k}{2}(\alpha + \beta_0)\right]^{2/3} - 1, \qquad (5.2.3)$$

$$(1 + kc)^{\frac{1}{2}} = a_1^{1/3} + a_2^{1/3}, \qquad (5.2.4)$$

$$\omega = \frac{2a_3}{k} - \frac{2}{3k}(a_1^{1/3} + a_2^{1/3}) = \frac{\alpha + \beta_0}{2} - \frac{1}{k}\left[\frac{k(\alpha + \beta_0)}{2}\right]^{1/3}, \qquad (5.2.5)$$

$$u = \frac{x}{t} - \omega = \alpha - \beta_0, \qquad (5.2.6)$$

$$\frac{x}{t} = \frac{3\alpha}{2} - \frac{\beta_0}{2} - \frac{1}{k}\left[\frac{k(\alpha + \beta_0)}{2}\right]^{1/3}. \qquad (5.2.7)$$

In the conventional gas dynamic case, problems involving the interaction of simple waves may be reduced to finding solutions of a single second order linear partial differential equation, the Euler-Poisson equation, for which the Riemann function may be expressed in terms of hypergeometric functions; for a particular set of values of the adiabatic index, the general solution may be written in terms of two arbitrary functions [43].

The existence of the relations given by Eqs. (5.2.3)—(5.2.6) raises the possibility of reducing problems involving the interaction of hydromagnetic simple waves to the solution of a single second order linear partial differential equation. This is carried out in the present section, but the results obtained are incomplete. In particular, it is shown that the requisite Riemann function may be found from a Fourier superposition of the solutions of a linear second order ordinary differential equation. Although the author has not been able to solve this latter equation in terms of the products of known functions, some circumstantial evidence is presented which makes it seem quite possible that better results may be obtained. However, it may well be the case that the Riemann function must be obtained numerically.

The second order partial differential equation is obtained by eliminating x from the Eqs. (1.2.6)—(1.2.7), i.e., from

$$x_\beta = (u + \omega) t_\beta$$
$$x_\alpha = (u - \omega) t_\alpha$$

and using the relations given by Eqs. (5.2.5)—(5.2.6). This gives

$$t_{\alpha\beta} + \frac{(9\phi^{2/3} - 1)(t_\alpha + t_\beta)}{6(\alpha + \beta)(\phi^{2/3} - 1)} = 0 \tag{5.2.8}$$

where $\phi = k(\alpha + \beta)/2$. It is convenient to introduce the new variable t^* by requiring that [44]

$$t_\alpha^* = x - (u + \omega) t ,$$
$$-t_\beta^* = x - (u - \omega) t .$$

Then $t^*(\alpha, \beta)$ satisfies the equation

$$t_{\alpha\beta}^* + \frac{(3\phi^{2/3} + 1)(t_\alpha^* + t_\beta^*)}{6(\alpha + \beta)(\phi^{2/3} - 1)} = 0 . \tag{5.2.9}$$

In order to transform Eq. (5.2.9) into a form for which results are known, a new dependent variable $l(\alpha, \beta)$ is introduced through the substitution

$$t^*(\alpha, \beta) = \frac{(\alpha + \beta)^{1/6} l(\alpha, \beta)}{\phi^{2/3} - 1} .$$

The resultant equation for l is

$$l_{\alpha\beta} + \frac{(9\phi^{2/3} + 7) l}{36(\alpha + \beta)^2 (\phi^{2/3} - 1)} = 0 . \tag{5.2.10}$$

The Riemann function may be obtained from the following result which was heuristically derived by RIEMANN [44] and proved rigorously by COHN [45].

Given the equation

$$l_{\alpha\beta} + H(\alpha + \beta) l = 0 , \tag{5.2.11}$$

let $\xi = \alpha + \beta$, $\eta = \alpha - \beta$ and

$$R^* = \frac{1}{\pi} \int_{-\infty}^{\infty} \exp[i\nu(\eta - \eta')] \, y(\xi, \xi'; \nu) \, d\nu$$

where $y(\xi, \xi'; \nu)$ is defined by the ordinary differential equation

$$\frac{d^2 y}{d\xi^2} + [\nu^2 + H(\xi)] y = 0 , \tag{5.2.12}$$

$$y = 0, \quad \frac{dy}{d\xi} = 1 \quad \text{for} \quad \xi = \xi'.$$

Then, changing back to the variables (α, β), R^* satisfies the relation

$$R^* = \frac{1}{2}\left[\text{sgn}\,(\alpha - \alpha') + \text{sgn}\,(\beta - \beta')\right] R\,(\alpha, \beta; \alpha', \beta')$$

where $R\,(\alpha, \beta; \alpha', \beta')$ is the Riemann function for Eq. (5.2.11).

From this result, the Riemann function of Eq. (5.2.10) may be obtained from the Fourier superposition of the solution of

$$\frac{d^2 y}{d\xi^2} + \left[\nu^2 + \frac{9 r_3 \xi^{2/3} + 7}{36 \xi^2 (r_3 \xi^{2/3} - 1)}\right] y = 0 , \qquad (5.2.13)$$

where $r_3 = (k/2)^{2/3}$, subject to the subsidiary conditions appended to Eq. (5.2.12). In order to put Eq. (5.2.13) into a form more convenient for comparison with known equations, let $z = \xi^{2/3}$, $y\,(\xi) = y_1\,(z)$ so that

$$z^2 \frac{d^2 y_1}{dz^2} - \frac{z}{2}\frac{dy_1}{dz} + \left[\frac{9\,z^3 \nu^2}{4} + \frac{9 r_3 z + 7}{16\,(r_3\,z - 1)}\right] y_1 = 0 .$$

Then let $r_3 z = \zeta$, $y_1\,(z) = y_2\,(\zeta)$ so that

$$\zeta^2 \frac{d^2 y_2}{d\zeta^2} - \frac{\zeta}{2}\frac{dy_2}{d\zeta} + \left[\frac{9}{4}\frac{\zeta^2}{r_3^3}\nu^2 + \frac{9\,(\zeta + 7/9)}{16\,(\zeta - 1)}\right] y_2 = 0 .$$

Since the smaller of the exponents relative to the singular points $(0, 1)$ is $\left(-\frac{1}{4}, 0\right)$, the substitution $y_2\,(\zeta) = \zeta^{-\frac{1}{4}} y_3\,(\zeta)$ is made which gives

$$\zeta\,(\zeta - 1)\,y_3''\,(\zeta) + (1 - \zeta)\,y_3'\,(\zeta) + \left[1 + \frac{9}{4}\frac{\nu^2 \zeta^2 (\zeta - 1)}{r_3^3}\right] y_3\,(\zeta) = 0 . \qquad (5.2.14)$$

The normal form of Eq. (5.2.14) is obtained by making the substitution $y_3\,(\zeta) = \zeta^{\frac{1}{2}} y_4\,(\zeta)$; this gives

$$\frac{d^2 y_4}{d\zeta^2} + \left[-\frac{3}{4\zeta^2} + \frac{1}{\zeta\,(\zeta - 1)} + \frac{9}{4}\frac{\nu^2 \zeta}{r_3^3}\right] y_4 = 0 . \qquad (5.2.15)$$

It has not been possible to relate Eqs. (5.2.14)–(5.2.15) to any known equations. But the following observations are somewhat suggestive. For $\nu = 0$, Eq. (5.2.14) reduces to the hypergeometric equation, and for very large ζ, Eq. (5.2.14) is close to one which has Airy functions as solutions. Since, in addition, the conventional gas dynamic case is governed by an equation whose Riemann function may be given in terms of hypergeometric functions, there exists the possibility that better results might be obtainable. The author has been unable to make further progress along these lines, but more research is indicated.

5.3 Perturbation of a Centered Simple Wave Flow

For an initially simple wave flow, the perturbation equations are Eqs. (1.2.28)–(1.2.29). For convenience, these equations are repeated here, specialized to the case of a uniform duct.

$$R_t + (u_0 + \omega_0) R_x + \left\{\left[\frac{3\omega_0^2 + (\gamma - 2) c_0^2}{2\omega_0^2}\right] \times \right.$$
$$\left. \times R + \left[\frac{(\gamma - 2) c_0^2 - \omega_0^2}{2\omega_0^2}\right] S\right\} \alpha_x = \frac{T_0}{2}, \quad (5.3.1)$$

$$S_t + (u_0 - \omega_0) S_x + \left\{\left[\frac{\omega_0^2 + (2 - \gamma) c_0^2}{2\omega_0^2}\right] \times \right.$$
$$\left. \times R - \left[\frac{3\omega_0^2 + (\gamma - 2) c_0^2}{2\omega_0^2}\right] S\right\} \beta_x = -\frac{T_0}{2}. \quad (5.3.2)$$

For the case of no applied field, Eqs. (5.3.1)—(5.3.2) reduce exactly to the basic perturbation equations derived in [2].

In the present section, it is assumed that the wave is centered at the origin and characterized by $\beta = \beta_0$, a constant. For convenience, the subscript zero will be dropped on the base flow. On each characteristic $dx/dt = u + \omega, (u, c, B)$ are constant, and these characteristics are straight lines in the (x, t) plane so that the wave may be represented as:

$$x = (u + \omega)t \quad (5.3.3)$$
$$\beta = \beta_0. \quad (5.3.4)$$

From the definition of the characteristic parameters of the base flow, it follows that

$$u = \alpha - \beta_0 \quad (5.3.5)$$
$$w = (\gamma - 1) (\alpha + \beta_0)/2. \quad (5.3.6)$$

In Eq. (5.3.1), an expression for α_x is needed, and this is obtained by differentiating Eq. (5.3.3) with respect to x which gives the result

$$\frac{1}{t} = u_x + \omega_x. \quad (5.3.7)$$

However, since $\omega_x = \left[\frac{\omega^2 + (\gamma - 2) c^2}{(\gamma - 1) \omega^2}\right] w_x$, Eq. (5.3.7) may be written as

$$\frac{1}{t} = u_x + \left[\frac{\omega^2 + (\gamma - 2) c^2}{(\gamma - 1) \omega^2}\right] w_x. \quad (5.3.8)$$

The derivatives u_x and w_x may be expressed in terms of α_x through Eqs. (5.3.5)—(5.3.6) so that

$$\alpha_x = 2\omega^2/[3\omega^2 + (\gamma - 2) c^2]t. \quad (5.3.9)$$

Thus, for the wave centered at the origin, Eqs. (5.3.1)—(5.3.2) reduce to

$$R_t + (u + \omega) R_x + \left\{R + \left[\frac{(\gamma - 2) c^2 - \omega^2}{3\omega^2 + (\gamma - 2) c^2}\right] S\right\} \frac{1}{t} = \frac{T_0}{2}, \quad (5.3.10)$$

$$S_t + (u - \omega) S_x = -\frac{T_0}{2}. \quad (5.3.11)$$

Eqs. (5.3.10) and (5.3.11) will be solved by considering the homogeneous system with $T_0 = 0$, and the physical significance of this reduced system is that to the order of terms retained, the perturbed flow is still isentropic, and then the solution of the complete system may be obtained by adding a particular solution of the complete system. The function S may be obtained from Eq. (5.3.11), and then R may be determined from Eq. (5.3.10).

The characteristics of the homogeneous equation associated with Eq. (5.3.11) are

$$dS = 0 \quad \text{and} \quad dx/(u - \omega) = dt . \tag{5.3.12}$$

From Eq. (5.3.3), the second of Eqs. (5.3.12) may be written as

$$\frac{dx}{dt} = u - \omega = u + \omega + t \frac{d}{dt} (u + \omega) . \tag{5.3.13}$$

From the definitions of the characteristic parameters and ω, it follows that

$$u + \omega = \frac{2}{(\gamma - 1)} \int [1 + k c^{\eta*}]^{\frac{1}{2}} dc - 2\beta_0 + c [1 + k c^{\eta*}]^{\frac{1}{2}} ,$$

so that Eq. (5.3.13) may be written as

$$2\omega + t \left\{ \frac{(\gamma + 1)[1 + k c^{\eta*}]^{\frac{1}{2}}}{\gamma - 1} + \frac{k \eta c^{\eta*}}{2[1 + k c^{\eta*}]^{\frac{1}{2}}} \right\} \frac{dc}{dt} = 0$$

which has the solution

$$t^2 c^{\frac{\gamma+1}{\gamma-1}} [1 + k c^{\eta*}]^{\frac{1}{2}} = \text{cst}$$

or

$$t^2 \omega c^{\frac{2}{\gamma-1}} = \text{cst} \tag{5.3.14}$$

or

$$\varrho \omega t^2 = \text{cst} . \tag{5.3.15}$$

Eq. (5.3.15) gives a convenient representation of the curvilinear cross-characteristics of the simple wave. Since the two first integrals obtained from Eq. (5.3.12) are $S = $ constant and Eq. (5.3.15), a convenient form for the solution of the homogeneous equation associated with Eq. (5.3.1) is

$$S = 2(\varrho \omega)^{\frac{1}{2}} t F' [\varrho \omega t^2] \tag{5.3.16}$$

where F is an arbitrary differentiable function. This form of solution is chosen in order to obtain a simpler form for the solution for R.

By the use of Eq. (5.3.3), it is seen that the homogeneous equation associated with Eq. (5.3.10) may be written as

$$t R_t + x R_x + R + \left[\frac{(\gamma - 2) c^2 - \omega^2}{3\omega^2 + (\gamma - 2) c^2} \right] S = 0 . \tag{5.3.17}$$

The solution for R may be obtained from the following system

$$\frac{dt}{t} = \frac{dx}{x} = \frac{dR}{-R + \left[\frac{\omega^2 + (2 - \gamma) c^2}{3\omega^2 + (\gamma - 2) c^2} \right] S} . \tag{5.3.18}$$

From the first two ratios of Eq. (5.3.18), one first integral is

$$x/t = \text{constant} \qquad (5.3.19)$$

i.e., the rectilinear characteristics. From the first and last ratios of Eq. (5.3.18)

$$\frac{d(Rt)}{dt} = \left[\frac{\omega^2 + (2-\gamma)c^2}{3\omega^2 + (\gamma-2)c^2}\right] \frac{d}{dt}\left[\frac{F(\varrho\omega t^2)}{(\varrho\omega)^{\frac{1}{2}}}\right]. \qquad (5.3.20)$$

To integrate Eq. (5.3.20), the first integral, Eq. (5.3.19) is utilized, i.e., on $x/t = \text{constant}$, ω, ϱ, c are constants, so that the solution of Eq. (5.3.20) may be written as:

$$Rt - \left[\frac{\omega^2 + (2-\gamma)c^2}{3\omega^2 + (\gamma-2)c^2}\right] \frac{F[\varrho\omega t^2]}{(\varrho\omega)^{\frac{1}{2}}} = \text{constant}. \qquad (5.3.21)$$

From the two first integrals, Eqs. (5.3.19) and (5.3.21), the solution of the homogeneous equation associated with Eq. (5.3.10) may be written as

$$Rt = \left[\frac{\omega^2 + (2-\gamma)c^2}{3\omega^2 + (\gamma-2)c^2}\right] \frac{F[\varrho\omega t^2]}{(\varrho\omega)^{\frac{1}{2}}} + G\left(\frac{x}{t}\right) \qquad (5.3.22)$$

where G is an arbitrary function. Thus the isentropic perturbation of a centered simple wave is given by Eqs. (5.3.16) and (5.3.22), and the following result is immediately apparent by comparison with [2]:

In the limit of vanishing magnetic field, the isentropic perturbation of a centered magnetohydrodynamic simple wave reduces precisely to the solution of the corresponding problem for conventional non-conducting fluids.

Since the general solution to the homogeneous system associated with Eqs. (5.3.10)—(5.3.11) has been found, the case of non-isentropic perturbed flow is reduced to finding a *particular* solution of Eqs. (5.3.10)—(5.3.11). The solution may be accomplished quite readily by choosing a convenient form for the function T_0.

The particle paths in the simple wave are given by $\varrho\omega t = \text{constant}$ or, more conveniently by

$$y = [1 + kc^{\eta*}]^{\frac{1}{4}} c^{(\gamma+1)/2\,(\gamma-1)} t^{\frac{1}{2}} = \text{constant}. \qquad (5.3.23)$$

Then from the definition of T_0,

$$T_0 = \frac{\varrho c^2 \Gamma'(\Psi_0)}{\gamma(\gamma-1)c_v} = \frac{\varrho\omega t\, c^2}{\gamma(\gamma-1)c_v\,\omega t} \Gamma'(\Psi_0)$$

so that it is convenient to write

$$T_0 = \frac{2c^2 \Omega'(y)}{\omega t} \qquad (5.3.24)$$

where Ω is an arbitrary differentiable function and y is given by Eq. (5.2.23). Then the particular solution for S is obtained from the equation

$$S_t + (u - \omega)S_x = -\frac{c\Omega'(y)}{(1+kc^{\eta*})^{\frac{1}{2}}t},$$

i.e., from the ratios:

$$\frac{dt}{1} = \frac{dx}{u - \omega} = \frac{dS}{-c\,\Omega'(y)/t(1 + kc\eta^*)^{\frac{1}{2}}}. \tag{5.3.25}$$

From the first two ratios of Eq. (5.3.25), one first integral is

$$t^{\frac{1}{2}} c^{(\gamma+1)/4\,(\gamma-1)} [1 + kc\eta^*]^{1/8} = a \tag{5.3.26}$$

where a is constant. From the first and last ratios of Eq. (5.3.25)

$$\frac{dS}{dt} = - \frac{c\,\Omega' \left[c^{(\gamma+1)/2\,(\gamma-1)} (1 + kc\eta^*)^{\frac{1}{2}} t^{\frac{1}{2}} \right]}{(1 + kc\eta^*)^{\frac{1}{2}} t}. \tag{5.3.27}$$

Using the first integral, Eq. (5.3.26), and the definition of y, Eq. (5.3.27) may be written as:

$$\frac{dS}{dt} = \frac{2}{a^6} c^{\frac{2\gamma}{\gamma-1}} t^{5/2} \frac{d\,\Omega\,(a^2/t^{\frac{1}{2}})}{dt}. \tag{5.3.28}$$

Integrating Eq. (5.3.28) and performing an integration by parts gives

$$S = 2c^{\frac{2\gamma}{\gamma-1}} t^{5/2} \Omega\,[a^2/t^{\frac{1}{2}}]/a^6 -$$

$$- \frac{2}{a^6} \left\{ \int \Omega\,[a^2/t^{\frac{1}{2}}]\, d\left[c^{\frac{2\gamma}{\gamma-1}} t^{5/2} \right] \right\}_a + \text{constant}. \tag{5.3.29}$$

From the two first integrals, Eqs. (5.3.26) and (5.3.29), the *particular* solution for S may be written as:

$$S = \frac{2c^2\,\Omega\,(y)}{\omega y} - \frac{2}{\omega y\, t^{5/2}\, c^{2/(\gamma-1)}} \times$$

$$\times \left\{ \int \Omega\,[a^2/t^{\frac{1}{2}}]\, d[c^{2\gamma/(\gamma-1)}\, t^{5/2}] \right\}_a \equiv \frac{2c^2\,\Omega\,(y)}{\omega y} + \Omega_1(x, t) \tag{5.3.30}$$

where the notation $\{\ \}_a$ denotes that the quadrature is to be carried out with constant a which must be then replaced by its equivalent from Eq. (5.3.26) to give the particular solution for S. The quadrature is abbreviated by $\Omega_1(x, t)$.

The particular solution for R is obtained from the equation

$$tR_t + xR_x + \left[R + \left\{ \frac{(\gamma - 2)c^2 - \omega^2}{3\omega^2 + (\gamma - 2)c^2} \right\} S \right] = \frac{c^2\,\Omega'(y)}{\omega},$$

i.e., from the ratios:

$$\frac{dt}{t} = \frac{dx}{x} = \frac{dR}{-R + \left\{ \dfrac{\omega^2 + (2 - \gamma)c^2}{3\omega^2 + (\gamma - 2)c^2} \right\} S + \dfrac{c^2\,\Omega' \left[t^{\frac{1}{2}} c^{(\gamma+1)/2\,(\gamma-1)} \{1 + kc\eta^*\}^{\frac{1}{4}} \right]}{\omega}}.$$

From the first two ratios, one first integral is

$$x/t = h \tag{5.3.31}$$

where h is a constant. From the first and last ratios, the following differential equation is obtained:

$$\frac{d[Rt]}{dt} = \left[\frac{\omega^2 + (2-\gamma)c^2}{3\omega^2 + (\gamma-2)c^2}\right]\left[\frac{2c^2}{\omega y}\,\Omega(y) + \Omega_1(x,t)\right] +$$

$$+ \frac{2c^2 t^{\frac{1}{2}}}{\omega c^{(\gamma+1)/2\,(\gamma-1)}[1+kc\eta^*]^{\frac{1}{4}}}\frac{d\Omega}{dt}\,. \qquad (5.3.32)$$

To integrate Eq. (5.3.32), the first integral, Eq. (5.3.31) may be utilized, i.e., u and c may be treated as constants. This gives:

$$Rt = \left[\frac{\omega^2+(2-\gamma)c^2}{3\omega^2+(\gamma-2)c^2}\right]\int\Omega_1(ht,t)\,dt + \left[\frac{\omega^2+(2-\gamma)c^2}{3\omega^2+(\gamma-2)c^2}\right]\times$$

$$\times\frac{2c^2\int[\Omega(y)/t^{\frac{1}{2}}]\,dt}{\omega[1+kc\eta^*]^{\frac{1}{4}}c^{(\gamma+1)/2\,(\gamma-1)}} + \frac{2c^2\int t^{\frac{1}{2}}\frac{d\Omega}{dt}\,dt}{\omega c^{(\gamma+1)/2\,(\gamma-1)}[1+kc\eta^*]^{\frac{1}{4}}} + \text{constant}\,. \qquad (5.3.33)$$

Performing an integration by parts in the third term on the right-hand side of Eq. (5.3.33) and using this result and the first integral, Eq. (5.3.31), the final *particular* solution for R may be written as:

$$Rt = \frac{2c^2t\,\Omega(y)}{\omega y} + \frac{c^2t^{\frac{1}{2}}}{\omega y}\left[\frac{3(2-\gamma)c^2-\omega^2}{3\omega^2+(\gamma-2)c^2}\right]\left\{\int\frac{\Omega(y)\,dt}{t^{\frac{1}{2}}}\right\}_{h=x/t} +$$

$$+ \left[\frac{\omega^2+(2-\gamma)c^2}{3\omega^2+(\gamma-2)c^2}\right]\left\{\int\Omega_1(ht,t)\,dt\right\}_{h=x/t} \qquad (5.3.34)$$

where the notation $\{\ \}_{h=x/t}$ denotes that the quadrature is to be carried out with x/t constant; then, replacing h by x/t, the solution Eq. (5.3.34) is obtained.

The general solution for the non-isentropic perturbation of the centered simple wave is given by Eq. (5.3.16) plus Eq. (5.3.30) and Eq. (5.3.22) plus Eq. (5.3.34). For $\gamma = 5/3$, these results reduce to those presented in [30]. By comparing these results with those of [2], the following result may be stated:

In the limit of vanishing magnetic field, the non-isentropic perturbation of a centered magnetohydrodynamic simple wave does not reduce to the solution for the corresponding problem for non-conducting fluids.

5.4 Perturbation of an Arbitrary Simple Wave

In this section, the non-isentropic perturbation of a non-centered simple wave is determined by quadratures. Let the wave be characterized by $\beta = \beta_0$, a constant. Then dropping the subscript zero on the base flow and letting $[x_0(z), t_0(z)]$ be the parametric representation of a curvilinear characteristic of the wave, the wave may be represented by

$$x = x_0(z) + [u(z) + \omega(z)]\tau, \quad t = t_0(z) + \tau, \tag{5.4.1}$$

$$-\frac{u(z)}{2} + \frac{w(z)}{(\gamma - 1)} = \beta_0. \tag{5.4.2}$$

The parameter τ defined in Eq. (5.4.1) should not be confused with the use of this symbol as a shock strength in Chapters 2 and 3. This symbol is used here to agree with the notation of [2], [5], [6], [38], [39] and [40] and to facilitate comparison with these papers.

The governing perturbation equations are Eqs. (5.3.1)—(5.3.2), and because of Eq. (5.4.2), Eq. (5.3.2) reduces to

$$S_t + (u - \omega)S_x = -T_0/2. \tag{5.4.3}$$

In order to solve Eq. (5.4.3), it is convenient to change from the independent variables (x, t) to (z, τ). With the definition that $S(x, t) = S_1(z, \tau)$, it follows that:

$$J S_x = S_{1z} - t_0' S_{1\tau},$$

$$J S_t = [x_0' + \tau(u' + \omega')]S_{1\tau} - (u + \omega)S_{1z},$$

$$J = x_0' + \tau(u' + \omega') - (u + \omega)t_0'$$

where primes denote differentiation with respect to the argument.

From the definitions of the characteristic parameters and ω, it follows that:

$$u + \omega = 2w(z)/(\gamma - 1) - 2\beta_0 + c[1 + kc^{\eta^*}]^{\frac{1}{2}}$$

so that

$$u' + \omega' = \left[\frac{3\omega^2 + (\gamma - 2)c^2}{(\gamma - 1)\omega c}\right]c'. \tag{5.4.4}$$

Since (x_0, t_0) is a curvilinear characteristic, $x_0' = (u - \omega)t_0'$, so that by the use of Eq. (5.4.4), it follows that the Jacobian may be written as

$$J = -2\omega t_0' + \left[\frac{3\omega^2 + (\gamma - 2)c^2}{(\gamma - 1)\omega c}\right]\tau c'. \tag{5.4.5}$$

Under the prescribed transformation of independent variables, Eq. (5.4.3) becomes

$$-2\omega S_{1z} + \frac{[3\omega^2 + (\gamma - 2)c^2]c'\tau}{(\gamma - 1)\omega c}S_{1\tau} = -\frac{JT_0}{2}. \tag{5.4.6}$$

Since the particle paths in the simple wave are given by $\varrho\omega\tau = $ constant or, more conveniently by

$$y = c^{(\gamma+1)/2(\gamma-1)}[1 + kc^{\eta^*}]^{\frac{1}{4}}\tau^{\frac{1}{2}} = \text{constant} \tag{5.4.7}$$

and

$$T_0 = \varrho\omega\tau\frac{c^2\Gamma'(\Psi_0)}{\gamma(\gamma-1)c_v\omega\tau},$$

it is convenient to write

$$\frac{T_0}{2} = \frac{c^2 \Omega'(y)}{\omega \tau} \qquad (5.4.8)$$

where Ω is an arbitrary differentiable function. The solution of Eq. (5.4.6) may be obtained from the ratios:

$$\frac{dz}{-2\omega} = \frac{d\tau}{\left[\dfrac{3\omega^2 + (\gamma - 2)c^2}{(\gamma - 1)\omega c}\right]\tau c'} = \frac{dS_1}{-\dfrac{JT_0}{2}} . \qquad (5.4.9)$$

From the first two ratios of Eq. (5.4.9), the following first integral is obtained:

$$c^{(\gamma+1)/4\,(\gamma-1)} \tau^{\frac{1}{2}} [1 + kc^{\eta*}]^{1/8} = a \qquad (5.4.10)$$

a constant or

$$\varrho \omega \tau^2 = \text{constant} . \qquad (5.4.11)$$

Eq. (5.4.11) gives a convenient representation of the curvilinear cross-characteristics of the simple wave. From the first and third ratios of Eq. (5.4.9), the following differential equation is obtained:

$$2\,dS_1 = \left[\frac{3\omega^2 + (\gamma - 2)c^2}{(\gamma - 1)\omega^2(1 + kc\eta*)^{\frac{1}{2}}}\right]\Omega'(y)\,dc - \frac{2c^2\,\Omega'(y)\,dt_0}{\omega\tau} . \qquad (5.4.12)$$

In order to integrate Eq. (5.4.12), the previous first integral may be used, and this gives a relation between z and τ, viz.,

$$\frac{dz}{d\tau} = -\frac{2(1 + kc\eta*)(\gamma - 1)c}{[3kc\eta* + \gamma + 1]\tau c'} \qquad (5.4.13)$$

and from the definition of y:

$$y = a^2/\tau^{\frac{1}{2}} .$$

Consequently, Eq. (5.4.12) may be written as

$$\frac{a^6\,dS_1}{2} = \tau^{5/2}c^{2\gamma/(\gamma-1)}\,d\Omega + c^{2\gamma/(\gamma-1)}\tau^{5/2}\frac{d\Omega}{d\tau}\,dt_0 . \qquad (5.4.14)$$

Integrating Eq. (5.4.14), performing an integration by parts in the first term and introducing Eq. (5.4.13) in the second term gives

$$\frac{a^6}{2}S_1 = c^{2\gamma/(\gamma-1)}\,\tau^{5/2}\,\Omega(a^2/\tau^{\frac{1}{2}}) -$$

$$- \int \Omega(a^2/\tau^{\frac{1}{2}})\,d\,[c^{2\gamma/(\gamma-1)}\,\tau^{5/2}] -$$

$$- \int \frac{c^{2\gamma/(\gamma-1)}\,\tau^{5/2}\,2(\gamma-1)\omega^2\,ct_0'}{\tau[3\omega^2 + (\gamma-2)c^2]c'}\,d\Omega + \text{constant} .$$

Performing an integration by parts in the last integral gives the following first integral:

$$S_1 = 2c^{2\gamma/(\gamma-1)}\tau^{5/2}\Omega\left(a^2/\tau^{\frac{1}{2}}\right)/a^6 - \frac{2}{a^6}\left\{\int\Omega\left(a^2/\tau^{\frac{1}{2}}\right)d\left[c^{2\gamma/(\gamma-1)}\tau^{5/2}\right]\right\}_a -$$

$$- \frac{4(\gamma-1)c^{(3\gamma-1)/(\gamma-1)}\tau^{3/2}\omega^2 t_0'\,\Omega\left[a^2/\tau^{\frac{1}{2}}\right]}{a^6[3\omega^2+(\gamma-2)c^2]c'} + \frac{4(\gamma-1)}{a^6}\times \qquad (5.4.15)$$

$$\times\left\{\int\Omega\left(a^2/\tau^{\frac{1}{2}}\right)d\left[c^{(3\gamma-1)/(\gamma-1)}\tau^{3/2}\frac{t_0'}{c'}\frac{\omega^2}{[3\omega^2+(\gamma-2)c^2]}\right]\right\}_a + \text{constant}\,.$$

From the two first integrals, Eqs. (5.4.11) and (5.4.15), the solution of Eq. (5.4.3) may be written as

$$S_1 = 2(\varrho\omega)^{\frac{1}{2}}\tau F'\left[\varrho\omega\tau^2\right] + \frac{2c^2\Omega(y)}{\omega y\tau}\times$$

$$\times\left[\tau - \frac{2(\gamma-1)c\omega^2 t_0'}{[3\omega^2+(\gamma-2)c^2]c'}\right] - \frac{2}{\omega y\tau^{5/2}c^{2/(\gamma-1)}}\times$$

$$\times\left\{\int\Omega\left(a^2/\tau^{\frac{1}{2}}\right)d\left[\tau^{3/2}c^{2\gamma/(\gamma-1)}\left\{\tau - \frac{2(\gamma-1)c\omega^2 t_0'}{[3\omega^2+(\gamma-2)c^2]c'}\right\}\right]\right\}_a \equiv \qquad (5.4.16)$$

$$\equiv 2(\varrho\omega)^{\frac{1}{2}}\tau F'\left[\varrho\omega\tau^2\right] + \frac{2c^2\Omega(y)}{\omega y\tau}\left[\tau - \frac{2(\gamma-1)c\omega^2 t_0'}{[3\omega^2+(\gamma-2)c^2]c'}\right] + \Phi(z,\tau)$$

where F is an arbitrary differentiable function, $\{\ \}_a$ has the same significance as in Eq. (5.3.30) and Φ is used to denote the quadrature in order to simplify the notation.

From the definition of the characteristic parameters of the basic flow and Eq. (5.4.4), it follows that

$$\alpha_x = \frac{2\omega c'}{(\gamma-1)c}\,z_x = \frac{2\omega c'}{(\gamma-1)cJ}$$

$$= \frac{2\omega^2 c'}{c'\tau[3\omega^2+(\gamma-2)c^2]-2(\gamma-1)\omega^2 ct_0'}\,. \qquad (5.4.17)$$

Substitution of Eq. (5.4.17) into Eq. (5.3.1) gives the equation for R, viz.,

$$R_t + (u+\omega)R_x + \left\{R + \left[\frac{(\gamma-2)c^2-\omega^2}{3\omega^2+(\gamma-2)c^2}\right]S\right\}\times$$

$$\times\frac{1}{\left\{\tau - \frac{2(\gamma-1)\omega^2 ct_0'}{[3\omega^2+(\gamma-2)c^2]c'}\right\}} = \frac{T_0}{2}\,. \qquad (5.4.18)$$

The solution of Eq. (5.4.18) is obtained from the ratios

$$\frac{dt}{1} = \frac{dx}{u+\omega} = \frac{dR}{-\dfrac{R}{\nu}+\left[\dfrac{\omega^2+(2-\gamma)c^2}{3\omega^2+(\gamma-2)c^2}\right]\dfrac{S}{\nu}+\dfrac{T_0}{2}} \qquad (5.4.19)$$

where ν denotes $\tau - \dfrac{2(\gamma-1)\,\omega^2 ct_0'}{[3\omega^2+(\gamma-2)c^2]c'}$. The integral of the first two ratios of Eq. (5.4.19) has the rectilinear characteristics for level curves. This first integral may be written as:

$$z(x, t) = h, \quad \text{a constant} \tag{5.4.20}$$

where the function is defined implicitly by:

$$x - x_0(z) = [u(z) + \omega(z)] [t - t_0(z)] .$$

From the first and last ratios of Eq. (5.4.19), the following differential equation is obtained

$$\frac{d}{dt} [\nu R] = \left[\frac{\omega^2 + (2 - \gamma) c^2}{3\omega^2 + (\gamma - 2) c^2} \right] \frac{d}{dt} \left[\frac{F(\varrho\omega\tau^2)}{(\varrho\omega)^{\frac{1}{2}}} \right] + \left[\frac{\omega^2 + (2 - \gamma) c^2}{3\omega^2 + (\gamma - 2) c^2} \right] \times$$

$$\times \Phi(h, t - t_0) + \left[\frac{\omega^2 + (2 - \gamma) c^2}{3\omega^2 + (\gamma - 2) c^2} \right] \frac{2c^2 \nu \Omega(y)}{\omega y \tau} + \frac{c^2 \nu \Omega'(y)}{\omega \tau} . \tag{5.4.21}$$

In order to integrate Eq. (5.4.21), the first integral Eq. (5.4.20) may be used, i.e., z may be treated as a constant. This gives

$$\nu R = \left[\frac{\omega^2 + (2 - \gamma) c^2}{3\omega^2 + (\gamma - 2) c^2} \right] \frac{F[\varrho\omega\tau^2]}{(\varrho\omega)^{\frac{1}{2}}} + \left[\frac{\omega^2 + (2 - \gamma) c^2}{3\omega^2 + (\gamma - 2) c^2} \right] \frac{2c^2}{\omega} \int \times$$

$$\times \frac{\Omega(y)}{y(t - t_0)} \left[t - t_0 - \frac{2(\gamma - 1) c\omega^2 t_0'}{[3\omega^2 + (\gamma - 2) c^2] c'} \right] dt + \frac{c^2}{\omega} \int \frac{\Omega'(y)}{t - t_0} \times$$

$$\times \left[t - t_0 - \frac{2(\gamma - 1) c\omega^2 t_0'}{[3\omega^2 + (\gamma - 2) c^2] c'} \right] dt + \left[\frac{\omega^2 + (2 - \gamma) c^2}{3\omega^2 + (\gamma - 2) c^2} \right] \int \times$$

$$\times \Phi(h, t - t_0) dt + \text{constant} . \tag{5.4.22}$$

Since

$$\int \frac{\nu \Omega'(y) dt}{t - t_0} = \frac{2\Omega(y) \nu}{y} - 2 \int \Omega(y) d\left[\frac{\nu}{y} \right],$$

the following general solution is obtained from the first integrals, Eqs. (5.4.20) and (5.4.22)

$$\nu R = G[z] + \left[\frac{\omega^2 + (2 - \gamma) c^2}{3\omega^2 + (\gamma - 2) c^2} \right] \frac{F[\varrho\omega\tau^2]}{(\varrho\omega)^{\frac{1}{2}}} + \frac{2c^2 \nu \Omega(y)}{\omega y} +$$

$$+ \left\{ \left[\frac{\omega^2 + (2 - \gamma) c^2}{3\omega^2 + (\gamma - 2) c^2} \right] \frac{2c^2}{\omega} \int \frac{\Omega(y)}{y(t - t_0)} \nu dt +$$

$$+ \left[\frac{\omega^2 + (2 - \gamma) c^2}{3\omega^2 + (\gamma - 2) c^2} \right] \int \Phi(h, t - t_0) dt -$$

$$- \frac{2c^2}{\omega} \int \Omega(y) d(\nu/y) \right\}_{h = z} \tag{5.4.23}$$

where G is an arbitrary function and $\{ \}_{h = z}$ has the same significance as in Eq. (5.3.30).

Eqs. (5.4.16) and (5.4.23) give the general solution for the non-isentropic perturbation of an arbitrary simple wave in terms of three arbitrary functions F, G and Ω. For $\gamma = 5/3$, these results reduce to those previously presented in [40]. They also include the results of section 5.3 as a special case, viz., in order to obtain the solution for a wave centered at $(x, t) = (0, 0)$, it is sufficient to let $(x_0, t_0) = (0, 0)$ and $z = x/t$. In the limit of vanishing magnetic field, the solution of this section reduces to the solution for the corresponding problem in conventional gas dynamics [39] only for isentropic perturbed flow.

In the foregoing chapters, a method, for which the germane mathematical techniques were conveniently available, has been presented for discussing weakly non-isentropic quasi-one-dimensional magnetohydrodynamic flows subjected to a transverse magnetic field, and the basis was the close parallelism of the basic equations with those of conventional gas dynamics. The number of known solutions which were but special cases of the theory provided frequent authentication of its validity.

In conventional gas dynamics, simple waves are frequently useful for building up solutions of isentropic flow problems. Non-isentropic perturbations of simple waves may well provide convenient building blocks for non-isentropic flows. Further, it is well-known that the study of one-dimensional flows has led frequently to a better understanding of more significant physical problems and the same situation may be expected to obtain in magnetohydrodynamics. It is hoped that the techniques presented may prove useful in achieving this goal.

The chapter ends with a discussion of a particular class of non-isentropic flows, which does not fit exactly into the classification of any of the previous chapters, for which exact solutions are possible.

5.5 A Class of Exact Solutions of Non-Isentropic Flow

With a suitable choice of dependent variables, problems in one-dimensional unsteady gas dynamics may be reduced to finding solutions of a particular Monge-Ampère partial differential equation [48], [49]. If the particle paths and isobars coincide, this technique fails, so that an alternative procedure must be employed. This particular class of flows was the subject of a paper by WEIR [50], and it is the purpose of this section to show that WEIR's discussion may be extended to magnetohydrodynamic flows subjected to a transverse magnetic field and the resultant solutions reduce, in the limit of vanishing magnetic field, to those obtained by WEIR [51].

For this purpose, it is convenient to use (u, ϱ, P, B, s) as dependent variables, so that the governing equations are

$$\varrho_t + \varrho u_x + u \varrho_x = 0 , \qquad (5.5.1)$$

$$\varrho (u_t + u u_x) + P_x + B B_x/\mu = 0 , \qquad (5.5.2)$$

$$B_t + u B_x + B u_x = 0 , \qquad (5.5.3)$$

$$s_t + u s_x = 0 , \qquad (5.5.4)$$

$$P = \exp [(s - s^*)/c_v] \varrho^\gamma . \qquad (5.5.5)$$

Exactly as in Chapter 1, $\varrho\,[dx - u\,dt]$ is the exact differential of a function Ψ such that Ψ equated to a constant defines the particle paths, and, from its definition, Ψ satisfies the equation

$$\Psi_t + u\Psi_x = 0 . \tag{5.5.6}$$

If it is assumed that $P = P(\Psi)$, it follows from Eq. (5.5.6) that

$$P_t + u P_x = 0 . \tag{5.5.7}$$

Since P, ϱ and s are related by some equation of state, e.g., Eq. (5.5.5), Eqs. (5.5.4) and (5.5.7) give

$$\varrho_t + u \varrho_x = 0 \tag{5.5.8}$$

so that Eq. (5.5.1) reduces to

$$\varrho u_x = 0 \tag{5.5.9}$$

and $u = u(t)$ alone, say

$$u = -f'(t) . \tag{5.5.10}$$

From Eqs. (5.5.3) and (5.5.9), B satisfies the equation

$$B_t + u B_x = 0 . \tag{5.5.11}$$

Consequently, B, P, ϱ, s and Ψ satisfy the same partial differential equation. Further, no explicit assumption of an equation of state has been made in the derivation to this point, and this class of flows may be generated from the single assumption $B = B(\Psi)$.

When Eq. (5.5.10) is substituted into Eqs. (5.5.4), (5.5.6), (5.5.7), (5.5.8) and (5.5.11), it is seen that s, Ψ, P, ϱ and B are each functions of $y = x + f(t)$. Eq. (5.5.2) may be written as

$$-\varrho f''(t) + \frac{dP}{dy} + \frac{B}{\mu}\frac{dB}{dy} = 0 \tag{5.5.12}$$

and since B and P are constant along a trajectory, Eq. (5.5.12) reduces to

$$f''(t) = 2A$$

where A is constant, so that

$$f(t) = A t^2 + A_1 t + A_2 .$$

With no loss of generality A_2 may be taken as zero, and, if $u(0) = u_0$,

$$u(t) = -2A t + u_0$$

and the trajectories are given by

$$y = x + f(t) = x + A t^2 - u_0 t = \text{constant} , \tag{5.5.13}$$

i.e., a family of coaxial parabolas.

From the definition of Ψ, $d\Psi = \varrho\,dy$ so that from Eq. (5.5.2)

$$\varrho = \frac{1}{2A}\frac{d}{dy}\left[P + \frac{B^2}{2\mu}\right] \tag{5.5.14}$$

and since ϱ is constant along a trajectory, Eq. (5.5.14) may be integrated to give

$$P + \frac{B^2}{2\mu} = 2A\Psi + A_3$$

with some constant A_3. Consequently, there is the result:

There exists only one possible case in which $P = P(\Psi)$, or, equivalently, $B = B(\Psi)$. Therein, $P + B^2/2\mu$ is a linear function of Ψ; u is a linear function of t; the trajectories are parabolas.

For a polytropic gas [Eq. (5.5.5)]

$$c^2 = \frac{\gamma P}{\varrho} = \frac{2A\gamma P}{\frac{d}{dy}\left[P + \frac{B^2}{2\mu}\right]},$$

$$\omega^2 = \frac{B^2}{\mu\varrho} + \frac{2A\gamma P}{\frac{d}{dy}\left[P + \frac{B^2}{2\mu}\right]}.$$

(5.5.15)

For the particular class of flows under consideration, the characteristics are given by the trajectories and the curves $dx/dt = u \pm \omega$, i.e.,

$$\frac{dy}{dt} = \pm\omega$$

(5.5.16)

with the solutions

$$t = \pm \int \left\{ \frac{\frac{d}{dy}\left[P + \frac{B^2}{2\mu}\right]}{b^2 \frac{d}{dy}\left[P + \frac{B^2}{2\mu}\right] + 2A\gamma P} \right\}^{\frac{1}{2}} dy.$$

(5.5.17)

If the equation of state is prescribed, e.g., Eq. (5.5.5), and also one initial condition, e.g., $s = s(\Psi)$ or $P = P(x)$ at $t = 0$, the solution may be determined uniquely.

Example 1. Isentropic flow with $B = $ Constant. If, Eq. (5.5.5) is written as $P = K\varrho^\gamma$ with a constant K, it follows from Eq. (5.5.14) that

$$2Ay = \int \left[\frac{K}{P}\right]^{\frac{1}{\gamma}} dP$$

so that

$$P = \left[\frac{2A(\gamma - 1)y}{\gamma K^{1/\gamma}}\right]^{\frac{\gamma}{\gamma - 1}}.$$

(5.5.18)

From Eqs. (5.5.14) and (5.5.18)

$$\varrho = \left[\frac{2A(\gamma - 1)y}{\gamma K}\right]^{\frac{1}{\gamma - 1}}.$$

(5.5.19)

Consequently

$$c^2 = 2Ay,$$

$$\omega^2 = 2Ay + \frac{B^2}{\mu\varrho}$$

where ϱ is given by Eq. (5.5.19). Thus if it is assumed that $u_0 = 0$, $u = -2At$, $c^2 = 2A(x + At^2)$. The gas is initially at rest, and the solution corresponds to expansion into a vacuum.

Example 2. Assume the initial condition

$$\left[P + \frac{B^2}{2\mu}\right] = \left[P_0 + \frac{B_0^2}{2\mu}\right] \exp(\lambda_2 x)$$

with P_0, B_0 and λ_2 constants. It follows that $\left[P + \frac{B^2}{2\mu}\right] = \left[P_0 + \frac{B_0^2}{2\mu}\right] \times$ $\times \exp(\lambda_2 y)$ throughout the flow and, further,

$$\varrho = \frac{\lambda_2}{2A}\left[P_0 + \frac{B_0^2}{2\mu}\right]\exp[\lambda_2 y],$$

$$c^2 = \frac{2A\,\gamma\,P_0}{\lambda_2\left[P_0 + \frac{B_0^2}{2\mu}\right]},$$

$$\omega^2 = \frac{2A\left(\gamma P_0 + \frac{B_0^2}{\mu}\right)}{\lambda_2\left[P_0 + \frac{B_0^2}{2\mu}\right]}.$$

Thus, the speed of propagation of small disturbances is constant. If it is assumed that $u_0 = 0$, it follows that $u = -2At$ and the trajectories are given by $x + At^2 = $ constant. The characteristics are given by

$$x + At^2 = \pm\left\{\frac{\left[2A\frac{B_0^2}{\mu} + 2A\gamma P_0\right]}{\left[\lambda_2\left(P_0 + \frac{B_0^2}{2\mu}\right)\right]}\right\}^{\frac{1}{2}} t + \text{constant}.$$

Graphs of these characteristics are given in WEIR's paper. Finally, the entropy distribution is given by

$$s = c_v\lambda_2(1 - \gamma)y + \text{constant}.$$

In this example, the entropy gradient causes the sound speed to remain constant.

Example 3. Assume $P = \lambda_3 y$, $\frac{B^2}{2\mu} = \lambda_4 y$. Then

$$\varrho = \frac{(\lambda_3 + \lambda_4)}{2A},$$

$$c^2 = \frac{2A\gamma\lambda_3'y}{\lambda_3 + \lambda_4},$$

$$\omega^2 = \frac{2Ay(\lambda_3\gamma + 2\lambda_4)}{\lambda_3 + \lambda_4}$$

so that the density remains constant throughout the non-uniform flow.

5*

The characteristics are given by the trajectories and the curves

$$t = \pm \left[\frac{2(\lambda_3 + \lambda_4)(x + A t^2)}{A(2\lambda_4 + \gamma \lambda_3)} \right]^{\frac{1}{2}} + \lambda_5 \qquad (5.5.20)$$

where λ_5 is a constant of integration. Consequently, the two families of characteristics given by Eq. (5.5.20) are given by the same family of parabolas, viz.,

$$A \lambda_3(\gamma - 2) t^2 - 2\lambda_5(2\lambda_4 + \gamma \lambda_3) A t - 2(\lambda_3 + \lambda_4) x +$$
$$+ A(2\lambda_4 + \gamma \lambda_3) \lambda_5 = 0 . \qquad (5.5.21)$$

The envelope of Eq. (5.5.21) is $x + A t^2 = 0$, so that the conclusion in this case agrees with that obtained by WEIR, viz., Eq. (5.5.20) implies that the point at which any member of the family given by Eq. (5.5.21) touches the envelope is also the point which divides that part of a curve representing a characteristic of one family from that representing a characteristic of the other family. Since this example deals with a gas expanding into a vacuum and $x + A t^2 = 0$ is the particle path through the origin, it must be the front of the expanding gas. Consequently, this gives another example where a characteristic of one family meets a gas front at a tangent and is reflected off as a characteristic of the other family. This occurs because the slopes of the characteristics given by Eq. (5.5.20) are the same at the front since $\omega = 0$ there. A sketch of the characteristics is given in WEIR's paper. Finally, the entropy distribution is given by

$$s = c_v \log y + \text{constant} .$$

Chapter 6

Formation and Decay of Shock Waves

6.1 Introduction

In the previous five chapters, the close analogy between one-dimensional conventional gas dynamics and the particular class of hydromagnetic flows under consideration has been utilized to extend a rather large part of the linearized theory of the former to the magnetic case. The analogy may be exploited in yet another problem of interest, namely, the formation and decay of weak shock waves. This is due to the fact that in common with conventional gas dynamic shock waves, the entropy change across magnetohydrodynamic shock waves involves only terms of the third and higher order in the shock strength [1]. With this fact as a basis and because of the existence of generalized Riemann invariants and simple normal shock relations, it is possible to extend the Friedrichs theory

[46] of the formation and decay of shock waves to the magnetic case with little difficulty. Since the generalized Riemann invariants and an exact solution for simple wave flow are known explicitly for a monatomic fluid, the discussion will be limited to a monatomic fluid in order to simplify the expository analysis of this chapter.

Because the differences between a shock transition and a simple wave transition with the same strength and same initial state involve only terms of third and higher order in the strength, the Friedrichs theory, which may be used for an approximate description of the propagation of non-uniform shock waves which are not too strong, replaces the shock transition relations by those through a corresponding simple compression wave. Since simple waves exist only in isentropic flows, the replacement is equivalent to the assumptions that the entropy and an appropriate generalized Riemann invariant do not change across the shock. After the simple wave, which replaces the actual disturbance, the fluid must be again at rest. The method is most easily applied when the flow in front of the non-uniform shock wave is initially at rest, and, for simplicity, the present discussion will be limited to that case.

The critical evaluations of the Friedrichs theory which have been made will not be discussed, though, in this connection, see the paper by LIGHTHILL [47], who showed that although the Friedrichs theory neglected third order terms, some of the neglected terms can integrate to give a second order effect, particularly for very weak shock waves which take a very long time to decay. This occurs, for example, when discussing the tail shock of an N-wave, behind which the Friedrichs theory assumes undisturbed fluid while LIGHTHILL shows the region to be a compression wave.

6.2 The Simple Wave Transition

For a monatomic fluid, the generalized Riemann invariants are

$$\frac{u}{2} + \frac{(1 + kc)^{3/2}}{k} = \frac{u}{2} + \frac{1}{k}\left(\frac{\omega}{c}\right)^3 = \alpha, \tag{6.2.1}$$

$$-\frac{u}{2} + \frac{(1 + kc)^{3/2}}{k} = -\frac{u}{2} + \frac{1}{k}\left(\frac{\omega}{c}\right)^3 = \beta. \tag{6.2.2}$$

For a forward-facing simple wave, the basic relation is $\beta = \beta_0$ or

$$u - \frac{2}{k}(1 + kc)^{3/2} = u_0 - \frac{2}{k}(1 + kc_0)^{3/2} \tag{6.2.3}$$

where the subscript zero denotes quantities in the constant state in front of the simple wave. From Eq. (6.2.3)

$$c = -\frac{1}{k} + \frac{1}{k}\left[(1 + kc_0)^{3/2} + \frac{k}{2}(u - u_0)\right]^{2/3} \tag{6.2.4}$$

and since

$$\omega = c(1 + kc)^{1/2},$$

$$\frac{\varrho}{\varrho_0} = \left(\frac{c}{c_0}\right)^3,$$

$$\frac{P}{P_0} = \left(\frac{c}{c_0}\right)^5,$$

these quantities may be expressed in terms of $u - u_0$. The expansions of these quantities in terms of powers of $u - u_0$ are given by

$$c = c_0 + \frac{1}{3(1 + m_1^2)^{\frac{1}{2}}}(u - u_0) - \frac{m_1^2}{36 c_0 (1 + m_1^2)^2}(u - u_0)^2 + \cdots,$$

$$\varrho = \varrho_0 + \frac{\varrho_0}{c_0 (1 + m_1^2)^{\frac{1}{2}}}(u - u_0) + \frac{\varrho_0 (4 + 3 m_1^2)}{12 c_0^2 (1 + m_1^2)^2}(u - u_0)^2 + \cdots,$$

$$\frac{1}{\varrho} = \frac{1}{\varrho_0} - \frac{1}{\varrho_0 c_0 (1 + m_1^2)^{\frac{1}{2}}}(u - u_0) + \frac{(12 + 13 m_1^2)}{18 \varrho_0 c_0^2 (1 + m_1^2)^2}(u - u_0)^2 + \cdots,$$

$$\quad (6.2.5)$$

$$P = P_0 + \frac{\varrho_0 c_0}{(1 + m_1^2)^{\frac{1}{2}}}(u - u_0) + \frac{\varrho_0 (8 + 7 m_1^2)}{12 (1 + m_1^2)^2}(u - u_0)^2 + \cdots,$$

$$\omega = \omega_0 + \frac{(2 + 3 m_1^2)}{6 (1 + m_1^2)}(u - u_0) + \frac{m_1^2}{36 c_0 (1 + m_1^2)^{5/2}}(u - u_0)^2 + \cdots,$$

$$u + \omega = u_0 + \omega_0 + \frac{(8 + 9 m_1^2)}{6 (1 + m_1^2)}(u - u_0) + \frac{m_1^2}{36 c_0 (1 + m_1^2)^{5/2}}(u - u_0)^2 + \cdots.$$

In the limit of vanishing magnetic field, the expansions (6.2.5) reduce exactly to the corresponding expansions in the non-magnetic case.

For a forward-facing shock in a polytropic gas, these expansions are the same through the second order but differ in the third order and higher terms. For a shock of at most moderate strength, the Friedrichs theory uses these expansions as an approximation for the true expansions.

In order to obtain an expansion for the shock velocity, U, the exact shock relations are employed. For this purpose, it is convenient to use the relation

$$\left(\frac{U - u_0}{c_0}\right)^2 = \frac{6\sigma + m_1^2 \sigma (5 + \sigma)}{2 (4 - \sigma)}$$

which yields the expansion

$$U = u_0 + \omega_0 + \frac{8 + 9 m_1^2}{12 (1 + m_1^2)}(u - u_0) +$$

$$\quad (6.2.6)$$

$$+ \frac{69 m_1^4 + 136 m_1^2 + 64}{288 c_0 (1 + m_1^2)^{5/2}}(u - u_0)^2 + \cdots.$$

However, it is convenient to have an expansion of U in powers of $u + \omega - u_0 - \omega_0$. From Eq. (6.2.6) and the last of Eqs. (6.2.5), this expansion is

$$U = u_0 + \omega_0 + \frac{1}{2}(u + \omega - u_0 - \omega_0) + \frac{1}{2c_0(8 + 9m_1^2)^2} \times$$

$$\times \left[\frac{69m_1^4 + 136m_1^2 + 64}{4(1 + m_1^2)^{\frac{1}{2}}} - m_1^2(1 + m_1^2) \right] (u + \omega - u_0 - \omega_0)^2 + \cdots .$$

(6.2.7)

Exactly as in the non-magnetic case, which Eq. (6.2.7) includes as a special case, the following result obtains:

In first order, the velocity of a forward-facing shock wave is the mean value of the velocities of the forward-facing sound waves in front of, i.e., $(u_0 + \omega_0)$ and behind, i.e., $(u + \omega)$ the shock wave.

6.3 The Differential Equation for the Shock Path

In order to obtain a typical problem in which the motion of a non-uniform shock wave must be determined, consider an initially uniform forward-facing shock wave, propagating with constant speed into fluid at rest and let the shock be overtaken from behind by a forward-facing simple wave. See Fig. 8. In the interaction region, the shock will cease to be uniform and the subsequent flow will be non-isentropic. However, under the assumption that the shock is of at most moderate strength, the Friedrichs theory replaces the shock transition by a transition through a corresponding simple compression wave. Consequently, the flow behind the non-uniform shock is assumed to be isentropic and characterized by $\beta = \beta_0$, i.e., the region behind the shock is a simple wave, and, in fact, it is just the continuation of the original simple wave. In this way, it is possible to determine the flow behind the non-uniform shock independently of the shock. Then, the shock path may be fitted in by solving a linear ordinary differential equation.

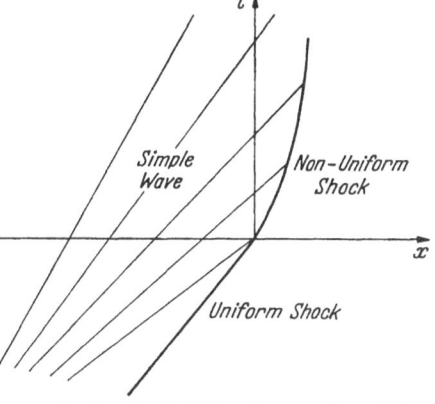

Fig. 8. At time $t = 0$ at the place $x = 0$, an initially uniform shock wave is overtaken and modified by a forward-facing simple wave

It is assumed that the fluid in front of the shock is a uniform state of rest, denoted by the subscript zero and that the interaction begins at $x = 0$ at $t = 0$. The simple wave behind the shock may be characterized by a characteristic parameter ξ, which is constant along each rectilinear characteristic, i.e., it is the value of x where the characteristic intersects the x-axis, and a function $\Omega(\xi)$. Specifically,

$$x = \xi + \Omega(\xi)t, \qquad (6.3.1)$$

$$\Omega(\xi) = u(\xi) + \omega(\xi),$$

$$-\frac{u(\xi)}{2} + \frac{1}{k}\left[\frac{\omega(\xi)}{c(\xi)}\right]^3 = \beta_0$$

and ξ is always negative.

It is possible to solve explicitly for $u(\xi)$ and $\omega(\xi)$ in terms of $\Omega(\xi)$. Actually, this is effectively the same problem as was solved in section 5.2 where an explicit solution was determined for the flow parameters in a centered simple wave. For the present problem, the previous solution, Eqs. (5.2.5)—(5.2.6), may be used. It is only necessary to replace x/t by Ω in those equations.

In order to determine the shock path, Eq. (6.3.1) is differentiated with respect to ξ. Then, since $dx = U\,dt$ along the shock path, the following linear ordinary differential equation for $t = t(\xi)$ is obtained

$$[U(\xi) - \Omega(\xi)]\frac{dt}{d\xi} - t\frac{d\Omega(\xi)}{d\xi} = 1 . \qquad (6.3.2)$$

From Eq. (6.2.7), U is a function of Ω and consequently of ξ; namely

$$\Omega(\xi) - U(\xi) = \frac{1}{2}\left[\Omega(\xi) - \omega_0\right] - \left[\Omega(\xi) - \omega_0\right]^2 \frac{1}{2c_0(8 + 9m_1^2)^2} \times$$

$$\times \left[\frac{69m_1^4 + 136m_1^2 + 64}{4(1 + m_1^2)^{\frac{1}{2}}} - m_1^2(1 + m_1^2)\right] . \qquad (6.3.3)$$

Inserting Eq. (6.3.3), which is correct through the second order, into Eq. (6.3.2), the solution may be determined uniquely when the initial condition $t = 0$ when $\xi = 0$ is appended. It is convenient to express the solution in terms of the function $\sigma_1(\xi)$ defined by

$$\Omega - \omega_0 = \omega_0\sigma_1(\xi) .$$

The explicit solution is

$$\omega_0 t(\xi) = 8\left[\frac{4 - D\sigma_1(\xi)}{\sigma_1(\xi)}\right]^2 \int_{\xi}^{0} \frac{\sigma_1(y)\,dy}{[4 - D\sigma_1(y)]^3} \qquad (6.3.4)$$

where

$$D = \frac{69m_1^4 + 136m_1^2 + 64 - 4m_1^2(1 + m_1^2)^{3/2}}{(8 + 9m_1^2)^2} .$$

Inserting Eq. (6.3.4) into the first of Eqs. (6.3.1) gives

$$x = \xi + 8[1 + \sigma_1(\xi)]\left[\frac{4 - D\sigma_1(\xi)}{\sigma_1(\xi)}\right]^2 \int_{\xi}^{0} \frac{\sigma_1(y)\,dy}{[4 - D\sigma_1(y)]^3} . \qquad (6.3.5)$$

Eqs. (6.3.4)—(6.3.5) give a parametric representention of the shock path.

In the limit of vanishing magnetic field, these results reduce precisely to those obtained by FRIEDRICHS, i.e., $\lim\limits_{m_1^2 \to 0} D = 1$. Further, the basic conclusion is effectively the same, i.e.,

For fixed m_1, and a compressive (expansive) simple wave, the shock front is accelerated (decelerated) and its strength increased (decreased).

It should be noted that Eqs. (6.3.4)—(6.3.5) are based on equations in which terms $0(\sigma_1^3)$ have been neglected. Consequently, terms $0(\sigma_1^3)$ must also be neglected in these equations.

6.4 Decaying Shock Wave

In the problem of section 6.3, suppose the interaction created when the shock wave is overtaken by a rarefaction wave continues indefinitely. From Eq. (6.3.4), $t \to \infty$ if $\sigma_1 \to 0$, i.e., $\Omega \to \omega_0$, which means that the slope of the last characteristic of the simple wave is ω_0, so that the gas velocity vanishes behind the rarefaction wave. Following FRIEDRICHS, the asymptotic behavior of the shock wave will be determined in the case that the pressure behind the rarefaction wave is equal to the pressure in front of the shock wave. Then, the gas velocity vanishes behind the tail of the rarefaction wave, and the tail is given by $\xi = \xi_0 < 0$ for which $\sigma_1(\xi_0) = 0$. Exactly as in the nonmagnetic case, it will be seen that the width of the wave zone, i.e., the distance between the shock front and the characteristic $\xi = \xi_0$ [$d(t)$ in Fig. 9] increases like $t^{\frac{1}{2}}$.

With the notation

$$A = 32 \int_{\xi_0}^{0} \frac{\sigma_1(y)\, dy}{[4 - D\sigma_1(y)]^3}$$

the following asymptotic expansion is obtained from Eq. (6.3.4)

$$(\omega_0 t)^{\frac{1}{2}} = 2A^{\frac{1}{2}}\left[\frac{1}{\sigma_1} - \frac{D}{4}\right]$$

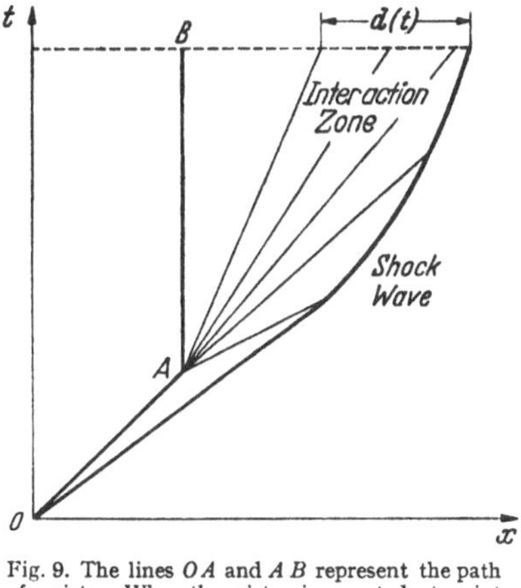

Fig. 9. The lines OA and AB represent the path of a piston. When the piston is arrested at point A, a simple rarefaction wave is generated. The simple wave catches up with and causes the initially uniform shock wave to decay. $d(t)$ is the width of the wave zone

which may be inverted to give

$$\sigma_1 = 2\left(\frac{A}{\omega_0 t}\right)^{\frac{1}{2}} - \frac{DA}{\omega_0 t} + \cdots . \tag{6.4.1}$$

Substituting Eq. (6.4.1) into Eq. (6.3.1) gives the asymptotic representation of the shock path, viz.,

$$x = \xi_0 + \omega_0 t + 2(A\,\omega_0 t)^{\frac{1}{2}} - AD .$$

Consequently, the width of the wave zone, $d(t)$, is given asymptotically by

$$d(t) = 2(A\,\omega_0 t)^{\frac{1}{2}} - A\,D\,. \tag{6.4.2}$$

The effect of the magnetic field on the width is contained entirely in the term D, and, in the limit of vanishing magnetic field, Eq. (6.4.2) reduces precisely to the corresponding result in the non-magnetic case. For small m_1, the quantity D decreases with increasing m_1, so that the asymptotic expression for the width of the wave zone increases with increasing m_1.

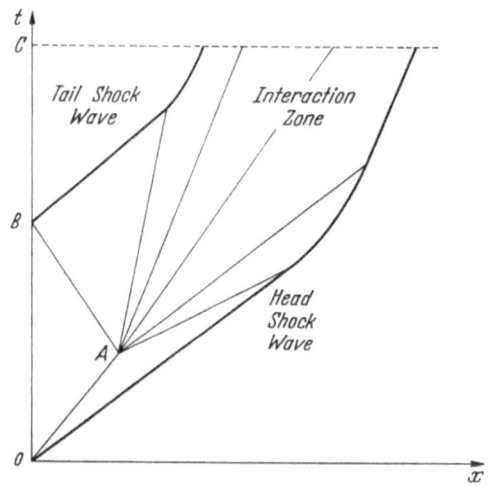

Fig. 10. The lines OA, AB and BC represent the path of a piston. The decaying N-wave may be discussed by the use of the Friedrichs theory

The asymptotic distribution of u, ω and P within the interaction region depends on the behavior of the original rarefaction wave near $x = \xi_0$, $t = 0$. Expanding $\sigma_1(\xi)$ near $\xi = \xi_0$

$$\sigma_1(\xi) = a_4(\xi - \xi_0) + a_5(\xi - \xi_0)^2 + \cdots . \tag{6.4.3}$$

Substitution of Eq. (6.4.3) into Eq. (6.3.1) gives

$$x - \xi_0 - \omega_0 t = (a_4\omega_0 t + 1)\,(\xi - \xi_0) + a_5\omega_0 t(\xi - \xi_0)^2 + \cdots$$

which may be inverted to yield

$$\xi - \xi_0 = \frac{x - \xi_0 - \omega_0 t}{a_4\omega_0 t + 1} - \frac{a_5\omega_0 t(x - \xi_0 - \omega_0 t)^3}{(a_4\omega_0 t + 1)^3} + \cdots . \tag{6.4.4}$$

Substitution of Eq. (6.4.4) into Eq. (6.4.3) gives

$$\sigma_1 = \frac{a_4(x - \xi_0 - \omega_0 t)}{a_4\omega_0 t + 1} + \frac{a_5(x - \xi_0 - \omega_0 t)^2}{(a_4\omega_0 t + 1)^3} + \cdots . \tag{6.4.5}$$

Since $\Omega = \omega_0(1 + \sigma_1(\xi))$, the asymptotic distribution of Ω in the interaction zone follows directly from Eq. (6.4.5). The asymptotic distributions for u and ω then follow directly from the explicit representations of these functions in terms of $\Omega(\xi)$.

From Eq. (6.4.5), it follows that

$$\sigma_1 = \frac{a_4 d(t)}{a_4 \omega_0 t + 1} + \frac{a_5 d^2(t)}{(a_4 \omega_0 t + 1)^3}$$

which is equivalent to Eq. (6.4.1). Exactly as in the non-magnetic case, it may be shown that the shock strength decreases like $t^{-\frac{1}{2}}$.

The problem of this section may be generated as illustrated in Fig. 9, and the constants a_4, a_5 and A may be determined quite easily [46].

Another problem, which may be treated in an entirely similar manner, is the decaying N-wave, see Fig. 10. In front of the head shock, the flow is known exactly, viz., a uniform state of rest, so that the flow behind the shock may be determined with an error of third or higher order in the shock strength, and the second order theory is valid for all time. However, the treatment of the tail shock presents difficulties since the flow which precedes it is only known approximately. See LIGHTHILL [47].

6.5 Formation of a Shock Wave

When a piston is pushed into a tube filled with gas at rest, a simple compression wave is generated, and, ultimately, the rectilinear characteristics, on each of which the flow parameters are constant, intersect and form an envelope. If the simple wave is generated by a uniformly accelerated piston motion and the initial acceleration is positive, the envelope begins at a point on the rectilinear characteristic through the origin, i.e., $x = \omega_0 t$, where the subscript zero denotes stagnation quantities. In this particular case, which is effectively the same as if the shock formed has been overtaken by the simple wave, the initial shock strength is zero and remains rather weak with a path close to the characteristic $x = \omega_0 t$ immediately after formation, so that the simplifying assumptions of the Friedrichs theory are satisfied.

Consequently, following FRIEDRICHS, it is assumed that the shock begins at $(x, t) = (0, 0)$ and the region in front of the shock is initially at rest. Thus, it is required that $\Omega(\xi_0) = \omega_0$, $d\Omega/d\xi < 0$ for $\xi < 0$ and $d\Omega/d\xi = -\infty$ at $\xi = 0$. Specifically, it is assumed that

$$\sigma_1(\xi) = (-\xi)^{\frac{1}{2}}[a + a_1\xi + \cdots], \qquad a > 0$$

with constants a and a_1. In order to obtain the shock path, Eqs. (6.3.1) and (6.3.4) are expanded with respect to powers of ξ, so that

$$\omega_0 t = \frac{4}{3a}(-\xi)^{\frac{1}{2}} - \frac{D\xi}{12},$$

$$x = \frac{4}{3a}(-\xi)^{\frac{1}{2}} - \frac{(4+D)\xi}{12}.$$

Consequently, the (x, t) representation of the shock path is

$$x = \omega_0 t + \frac{3}{16} a^2 \omega_0^2 t^2 + \cdots .$$

The shock path is initially close to the characteristic $x = \omega_0 t$, so that the initial shock velocity is ω_0 and the acceleration is $3 a^2 \omega_0^2 / 8$.

The discussion of this section may be particularized by specifiying the motion of the piston. The resultant simple compression wave is written in the general form

$$x = x_0(\lambda) + (u + \omega) [t - t_0(\lambda)] \tag{6.5.1}$$

where λ is a parameter. The piston is assumed to be pushed into the gas with constant acceleration until the time $t = t^*$, after which its velocity is maintained with the constant value given at that time. The fluid in front of the piston is assumed to be initially at rest with $u = u_0 = 0$, $c = c_0$, $\omega = \omega_0$. The specified piston motion is given by

$$x = \frac{r t^2}{2} \qquad\qquad 0 \le t \le t^*$$

$$x = r t^* (t - t^*) \qquad t \ge t^* .$$

Letting $\lambda = $ constant denote the rectilinear characteristic which originates at the piston at time $t = \lambda$, the compression simple wave, Eq. (6.5.1), may be represented by

$$x = \frac{r \lambda^2}{2} + (u + \omega) (t - \lambda)$$

$$-\frac{u}{2} + \frac{(1 + kc)^{3/2}}{k} = \beta_0 = \frac{1}{k} \left(\frac{\omega_0}{c_0} \right)^3 .$$

or

$$x = \frac{r \lambda^2}{2} + \left[\frac{3}{2} r \lambda + \beta_0 - \frac{1}{k} \left(k \beta_0 + \frac{k r \lambda}{2} \right)^{1/3} \right] (t - \lambda) . \tag{6.5.2}$$

Although the envelope of Eq. (6.5.2) is easily determined, say $[x^E(\lambda), t^E(\lambda)]$, the point at which the shock formation begins is determined by finding the minimum of $t^E(\lambda)$. Since this is given at best by a complicated implicit representation, it seems more fruitful to give an approximate representation for Eq. (6.5.2), valid for small applied fields. Since

$$\frac{1}{k} \left[k \beta_0 + \frac{k r \lambda}{2} \right]^{1/3} = \frac{(1 + kc_0)^{\frac{1}{2}}}{k} + \frac{r \lambda c_0^2}{6 \omega_0^2} -$$

$$- \frac{1}{36} \frac{k r^2 \lambda^2 c_0^5}{\omega_0^5} + O(k^2 r^3 \lambda^3) ,$$

the coefficient of $(t - \lambda)$ in Eq. (6.5.2) may be written as

$$\frac{3}{2} r \lambda + \omega_0 - \frac{r \lambda}{6 (1 + kc_0)} + \frac{k r^2 \lambda^2 c_0^5}{36 \omega_0^5} + O(k^2 r^3 \lambda^3)$$

and for simplicity, only the first three terms are retained. Then, the simple wave may be represented by

$$x = \frac{r\lambda^2}{2} + \left[\frac{3}{2}r\lambda + \omega_0 - \frac{r\lambda}{6(1+m_1^2)}\right](t-\lambda).$$ (6.5.3)

The envelope of Eq. (6.5.3) is given by

$$t^E = \frac{2(5+6m_1^2)\lambda}{(8+9m_1^2)} + \frac{6(1+m_1^2)^{3/2}c_0}{(8+9m_1^2)r},$$ (6.5.4)

$$x^E = \frac{r\lambda^2}{2} + \left\{\frac{(2+3m_1^2)\lambda}{(8+9m_1^2)} + \frac{6(1+m_1^2)^{3/2}c_0}{(8+9m_1^2)r}\right\} \times$$
$$\times \left\{\frac{3}{2}r\lambda + \omega_0 - \frac{r\lambda}{6(1+m_1^2)}\right\}.$$ (6.5.5)

The initial point of shock formation is given by

$$t^E_{\min} \equiv t_\nu = \frac{6(1+m_1^2)^{3/2}c_0}{(8+9m_1^2)r},$$

$$x_\nu = \omega_0 t_\nu.$$

Since the shock begins at (x_ν, t_ν) and lies within the region $x > \omega_0 t$, the region in front of the shock is a constant state so that the methods of this chapter are applicable.

By shifting the origin to the point (x_ν, t_ν), the simple wave may be represented by

$$x - x_\nu = (t - t_\nu)\left[\frac{(8+9m_1^2)r\lambda}{6(1+m_1^2)} + \omega_0\right] - \frac{(5+6m_1^2)r\lambda^2}{6(1+m_1^2)}.$$ (6.5.6)

Since $\Omega = \omega_0(1+\sigma_1) = \frac{(8+9m_1^2)r\lambda}{6(1+m_1^2)} + \omega_0$, it follows that

$$\sigma_1 = \frac{(8+9m_1^2)r\lambda}{6(1+m_1^2)\omega_0}$$ (6.5.7)

so that

$$\xi = -\frac{6(5+6m_1^2)(1+m_1^2)\omega_0^2\sigma_1^2}{r(8+9m_1^2)^2}.$$

In order to determine the motion of the shock wave, these results are substituted into Eq. (6.3.4). This gives

$$\omega_0(t-t_\nu) = 96\left[\frac{4-D\sigma_1}{\sigma_1}\right]^2 \frac{(5+6m_1^2)(1+m_1^2)\omega_0^2}{(8+9m_1^2)^2 r}\int_0^{\sigma_1}\frac{\sigma_1^2 d\sigma_1}{(4-D\sigma_1)^3}$$
$$= \frac{12(5+6m_1^2)(1+m_1^2)\omega_0^2}{(8+9m_1^2)^2 r}\left\{\frac{2}{3}\sigma_1 + \left(\frac{3}{8} - \frac{D}{3}\right)\sigma_1^2 + 0(\sigma_1^3)\right\}.$$ (6.5.8)

By inverting Eq. (6.5.8), the following expansion for σ_1 is obtained

$$\sigma_1 = \frac{(8+9m_1^2)^2(t-t_\nu)r}{8\omega_0(5+6m_1^2)(1+m_1^2)}\left[1 - \frac{(8+9m_1^2)^2(t-t_\nu)r}{128\omega_0(5+6m_1^2)(1+m_1^2)}\right].$$ (6.5.9)

Thus, from Eqs. (6.5.6), (6.5.7) and (6.5.9),

$$x - x_\nu = \omega_0(t-t_\nu) + \frac{r(8+9m_1^2)^2(t-t_\nu)^2}{32(5+6m_1^2)(1+m_1^2)} + \cdots.$$ (6.5.10)

In the limit of vanishing magnetic field, Eq. (6.5.10) reduces exactly to the result obtained by FRIEDRICHS for the non-magnetic case. The behavior of the other flow parameters in the simple wave may be obtained from their explicit representations and the above expansions.

The objectives of this chapter have been achieved, and it has been shown that the Friedrichs theory may be extended to the magnetic case with little difficulty.

Chapter 7

The Effects Due to an Oblique Applied Field

7.1 The Characteristic Form of the Governing Equations

The analysis of the foregoing chapters was restricted to the case of a transverse magnetic field, and the basis for the results was the striking parallelism between this particular class of hydromagnetic flows and conventional gas dynamic flows. For an arbitrary orientation of the magnetic field, magnetohydrodynamic wave phenomena are much more complicated, and, in fact, there exist three sound speeds, i.e., modes of propagation, called fast, slow and intermediate waves, in each direction which, moreover, depend on the direction of propagation. The anisotropic nature of magnetohydrodynamic wave propagation was noted first by HERLOFSON [56] and by VAN DE HULST [57]. An introductory discussion may be found, for example, in the papers by FRIEDRICHS [1], GRAD [58] and SHERCLIFF [55].

Naturally, it would be nice if the results of the previous chapters could be extended directly to the case of an oblique field, and the purpose of this chapter is to present the preliminary analysis necessary for that extension. The equations governing one-dimensional unsteady flow, subjected to an oblique magnetic field, are considered. Throughout, it is assumed that all dependent variables are functions of only one space variable x and the time t, i.e., it is the projections of the characteristics from (x, y, t) space onto the (x, t) plane that are considered. The governing equations are written in characteristic form and simple wave flows are discussed. For initially uniform or simple wave flows, a reduction in the number of governing equations is possible in a manner quite similar to that used in Chapter 1. General solutions are obtained for the non-isentropic perturbation of an initially uniform flow, an initially centered simple wave flow and an initially arbitrary simple wave flow.

It is assumed that the velocity vector has only two components, viz., $(u_1, u_2, 0)$ and the induction is given by $\vec{B} = (B_1, B_2, 0)$. Consequently, the basic equations are

$$c_t + u_1 c_x + \frac{(\gamma - 1)}{2}\, c u_{1x} = 0 , \tag{7.1.1}$$

$$u_{1t} + u_1 u_{1x} + \frac{2 c c_x}{\gamma - 1} + b_2^2 \frac{B_{2x}}{B_2} - \frac{c^2 s_x}{\gamma(\gamma - 1) c_v} = 0 , \tag{7.1.2}$$

$$u_{2t} + u_1 u_{2x} - b_1^2 \frac{B_{2x}}{B_1} = 0 , \tag{7.1.3}$$

$$B_{2t} - B_1 u_{2x} + u_{1x} B_2 + u_1 B_{2x} = 0 , \tag{7.1.4}$$

$$s_t + u_1 s_x = 0 , \tag{7.1.5}$$

where $b_i^2 = B_i^2/\mu \varrho$ $(i = 1, 2)$.
As a consequence of MAXWELL's equations, there is the further condition that $B_1 = \text{constant}$.

The possible characteristic surfaces $\phi(x, t) = \text{constant}$ of the system of Eqs. (7.1.1)—(7.1.5) are given by solutions of the following first-order partial differential equations for ϕ:

$$[\phi_t + u_1 \phi_x]\, [\phi_t^4 + 4 u_1 \phi_x \phi_t^3 + (6 u_1^2 - b_1^2 - b_2^2 - c^2)\, \phi_x^2 \phi_t^2 +$$
$$+ \{4 u_1^3 - 2 u_1 (b_1^2 + b_2^2 + c^2)\, \phi_x^3 \phi_t\} + \tag{7.1.6}$$
$$+ \phi_x^4 \{u_1^2 (u_1^2 - b_1^2 - b_2^2) - c^2 (u_1^2 - b_1^2)\}] = 0 .$$

The first factor shows that one characteristic curve is given by $dx/dt = u_1$. If $\phi(x, t) = \text{constant}$ is characteristic, then $dx/dt = - \phi_t/\phi_x$, so that the introduction of $dx/dt = u_1 + a$ leads to the following algebraic equation for a:

$$a^4 - \omega^2 a^2 + b_1^2 c^2 = 0 \tag{7.1.7}$$

where $\omega^2 = c^2 + b_1^2 + b_2^2$. The larger and smaller of the roots $a > 0$ of Eq. (7.1.7) will be denoted by a_f (fast speed) and a_s (slow speed), respectively [1].

Consequently, introducing characteristic parameters $(\alpha, \beta, \xi, \eta, \zeta)$, the basic system of Eqs. (7.1.1)—(7.1.5) may be written in the following characteristic form [52]

$$x_\beta = (u_1 + a_f) t_\beta ,$$
$$x_\alpha = (u_1 - a_f) t_\alpha ,$$
$$x_\xi = u_1 t_\xi ,$$
$$x_\eta = (u_1 + a_s) t_\eta ,$$
$$x_\zeta = (u_1 - a_s) t_\zeta ,$$

$$a_f u_{1\beta} + \frac{2 c c_\beta}{(\gamma - 1)} + (a_f^2 - c^2)\frac{B_{2\beta}}{B_2} - \frac{B_1 (a_f^2 - c^2) u_{2\beta}}{a_f B_2} - \frac{c^2 s_\beta}{\gamma(\gamma - 1) c_v} = 0 , \tag{7.1.8}$$

$$- a_f u_{1\alpha} + \frac{2 c c_\alpha}{\gamma - 1} + (a_f^2 - c^2)\frac{B_{2\alpha}}{B_2} + \frac{B_1 (a_f^2 - c^2) u_{2\alpha}}{a_f B_2} - \frac{c^2 s_\alpha}{\gamma(\gamma - 1) c_v} = 0 , \tag{7.1.9}$$

$$a_s u_{1\eta} + \frac{2cc_\eta}{\gamma-1} + (a_s^2 - c^2)\frac{B_{2\eta}}{B_2} - \frac{B_1(a_s^2 - c^2)u_{2\eta}}{a_s B_2} - \frac{c^2 s_\eta}{\gamma(\gamma-1)c_v} = 0, \quad (7.1.10)$$

$$-a_s u_{1\zeta} + \frac{2cc_\zeta}{\gamma-1} + (a_s^2 - c^2)\frac{B_{2\zeta}}{B_2} + \frac{B_1(a_s^2 - c^2)u_{2\zeta}}{a_s B_2} - \frac{c^2 s_\zeta}{\gamma(\gamma-1)c_v} = 0, \quad (7.1.11)$$

$$s_\xi = 0. \quad (7.1.12)$$

For the case of a transverse field only, $B_1 = 0$, $a_s = 0$ and $a_f = \omega = [b_2^2 + c^2]^{\frac{1}{2}}$, and the above characteristic system reduces to that derived in Chapter 1, viz., Eqs. (1.2.6)−(1.2.12).

In the transverse field case, the particle paths were characteristics of multiplicity two. Three independent characteristic directions led to four independent characteristic forms of the governing equations, and a reduction in the number of equations was possible. A similar reduction is possible in the present case for simple wave flows.

Hydromagnetic simple wave flow was considered by FRIEDRICHS [1], who showed that the problem could be reduced to finding the solution of a single linear first-order differential equation, which could be solved explicitly for the case of a polytropic gas, and by BAZER [54], SHERCLIFF [55] and others. In principle, simple wave flow would be characterized by integrating one of Eqs. (7.1.8)−(7.1.11) and setting the result equal to a constant thus determining a Riemann invariant. In actuality, a better approach is to look for solutions constant along a single family of rectilinear characteristics, i.e., to determine the motion of a "phase" which moves with constant velocity. For example, consider a forward-facing wave characterized by

$$x = [u_1(\delta) + a(\delta)]t + \delta \quad (7.1.13)$$

where δ is a parameter. From Eq. (7.1.13), it follows that

$$\frac{\partial}{\partial x} = \frac{1}{1 + [u_1'(\delta) + a'(\delta)]t}\frac{\partial}{\partial \delta},$$

$$\frac{\partial}{\partial t} = \frac{-[u_1(\delta) + a(\delta)]}{1 + [u_1'(\delta) + a'(\delta)]t}\frac{\partial}{\partial \delta}.$$

Substituting these results into Eqs. (7.1.1)−(7.1.5) gives

$$-ac_\delta + \frac{(\gamma-1)}{2}cu_{1\delta} = 0,$$

$$-au_{1\delta} + \frac{2cc_\delta}{\gamma-1} + \frac{b_2^2}{B_2}B_{2\delta} - \frac{c^2 s_\delta}{\gamma(\gamma-1)c_v} = 0,$$

$$-au_{2\delta} - \frac{b_1^2}{B_1}B_{2\delta} = 0, \quad (7.1.14)$$

$$-aB_{2\delta} - B_1 u_{2\delta} + B_2 u_{1\delta} = 0,$$

$$-as_\delta = 0,$$

whose eliminant is Eq. (7.1.7). After some simple manipulation, the system of Eqs. (7.1.14) may be put into a more convenient form, e.g., for a forward-facing fast wave

$$-\frac{u_{1\alpha}}{2} + \frac{a_f}{\gamma-1}\frac{c_\alpha}{c} = 0 \,,$$

$$\frac{B_{2\alpha}}{B_2} + \frac{2}{\gamma-1}\frac{a_f^2}{(b_1^2 - a_f^2)}\frac{c_\alpha}{c} = 0 \,,$$

$$u_{2\alpha} + \frac{b_1^2 B_{2\alpha}}{a_f B_1} = 0 \,,$$

$$u_{1\alpha} + \frac{B_1(a_f^2 - b_1^2)u_{2\alpha}}{B_2 b_1^2} = 0$$

(7.1.15)

where δ has been replaced by α to agree with Eq. (7.1.9). When the first of Eqs. (7.1.15) is integrated and set equal to a constant, a generalized Riemann invariant is determined which is completely analogous to Eq. (1.2.9) to which it reduces when $B_1 \to 0$. The system of Eqs. (7.1.15) is completely equivalent to Eq. (7.1.9).

The system corresponding to Eqs. (7.1.15) for a backward-facing fast wave is

$$\frac{u_{1\beta}}{2} + \frac{a_f}{\gamma-1}\frac{c_\beta}{c} = 0 \,,$$

$$\frac{B_{2\beta}}{B_2} + \frac{2}{\gamma-1}\frac{a_f^2}{(b_1^2 - a_f^2)}\frac{c_\beta}{c} = 0 \,,$$

$$u_{2\beta} - \frac{b_1^2 B_{2\beta}}{a_f B_1} = 0 \,,$$

$$u_{1\beta} + \frac{B_1(a_f^2 - b_1^2)}{B_2 b_1^2} u_{2\beta} = 0$$

(7.1.16)

where the characteristic parameter β has been chosen to agree with Eq. (7.1.8).

From the two systems of Eqs. (7.1.15) and Eqs. (7.1.16), it follows that the following differential relations hold on *both* the forward-facing and backward-facing characteristics:

$$\frac{dB_2}{B_2} + \frac{2}{\gamma-1}\frac{a^2}{b_1^2 - a^2}\frac{dc}{c} = 0 \,,$$

$$du_1 + \frac{B_1(a^2 - b_1^2)}{B_2 b_1^2} du_2 = 0$$

where the symbol a has been written without subscript since the same relations hold for $a = a_f$ or $a = a_s$. It follows immediately that these differential relations must hold throughout the flow, so that the following integrals are obtained:

$$\log B_2 + \frac{2}{\gamma-1}\int \frac{a^2 \, dc}{(b_1^2 - a^2)c} = \text{constant} \,,$$

(7.1.17)

$$u_2 + \frac{1}{B_1}\int \frac{B_2 b_1^2}{(a^2 - b_1^2)} du_1 = \text{constant} \,.$$

(7.1.18)

These two integrals may be used to reduce to number of equations governing the flow, viz., because of Eq. (7.1.17), it follows that Eq. (7.1.4) is identical to Eq. (7.1.1) and because of Eq. (7.1.18), it follows that Eq. (7.1.3) is identical to Eq. (7.1.2) when the entropy is constant. Consequently, the governing equations are reduced to

$$\frac{c_t}{\gamma-1} + \frac{u_1 c_x}{\gamma-1} + \frac{c u_{1x}}{2} + \frac{u_1 c A_x}{2A} = 0 , \qquad (7.1.19)$$

$$u_{1t} + u_1 u_{1x} + \frac{2a^2 c_x}{(\gamma-1)c} - \frac{c^2 s_x}{\gamma(\gamma-1)c_v} = 0 , \qquad (7.1.20)$$

$$s_t + u_1 s_x = 0 \qquad (7.1.21)$$

where an arbitrary, time-independent area variation has been added, and it should be noted that the system of Eqs. (7.1.19)—(7.1.21) is valid for isentropic flow only. The entropy has been included since these equations will be linearized in the neighborhood of a known isentropic state. In order to make the generalized Riemann invariants linear relations between the dependent variables, the variable w is introduced by the substitution $w = \int \frac{a}{c} dc$. Then, appropriate combinations of the resultant system lead to the following characteristic form of the equations

$$\left[\frac{u_1}{2} + \frac{w}{\gamma-1}\right]_t + (u_1 + a)\left[\frac{u_1}{2} + \frac{w}{\gamma-1}\right]_x$$
$$= -\frac{u_1 a A_x}{2A} + \frac{c^2 s_x}{2\gamma(\gamma-1)c_v} , \qquad (7.1.22)$$

$$\left[-\frac{u_1}{2} + \frac{w}{\gamma-1}\right]_t + (u_1 - a)\left[-\frac{u_1}{2} + \frac{w}{\gamma-1}\right]_x$$
$$= -\frac{u_1 a A_x}{2A} - \frac{c^2 s_x}{2\gamma(\gamma-1)c_v} , \qquad (7.1.23)$$

$$s_t + u_1 s_x = 0 . \qquad (7.1.24)$$

Then, linearizing Eqs. (7.1.22)—(7.1.24) in the neighborhood of a known (isentropic) constant area flow, i.e., a uniform or simple wave flow, denoted by the subscript zero, the following system of linear equations is obtained for the terms of first order, denoted by a bar:

$$R_t + (u_{10} + a_0) R_x + (\overline{u_1} + \bar{a})\alpha_x$$
$$= -\frac{u_{10} a_0 \overline{A}_x}{2A_0} + \frac{c_0^2 \overline{s}_x}{2\gamma(\gamma-1)c_v} , \qquad (7.1.25)$$

$$S_t + (u_{10} - a_0) S_x + (\overline{u_1} - \bar{a})\beta_x$$
$$= -\frac{u_{10} a_0 \overline{A}(x)}{2A_0} - \frac{c_0^2 \overline{s}_x}{2\gamma(\gamma-1)c_v} , \qquad (7.1.26)$$

$$\bar{s}_t + u_{10} \bar{s}_x = 0 \qquad (7.1.27)$$

where the first-order generalized Riemann invariants have been denoted by R and S, i.e.,

$$R = \frac{\bar{u}_1}{2} + \frac{\bar{w}}{\gamma - 1}, \qquad S = -\frac{\bar{u}_1}{2} + \frac{\bar{w}}{\gamma - 1}$$

and it has been noted that

$$\frac{u_{10}}{2} + \frac{w_0}{\gamma - 1} = \alpha, \qquad -\frac{u_{10}}{2} + \frac{w_0}{\gamma - 1} = \beta$$

are the characteristic parameters of the base flow. Throughout, the symbol a will appear without the subscript s or f since the analysis is valid for slow or fast waves, and the appropriate characteristic parameters will be denoted by (α, β). Further, the subscript zero will be dropped from base flow quantities in order to simplify the notation.

Since Eq. (7.1.7) for the wave speeds may be written in the equivalent form

$$\frac{b_2^2}{a^2 - b_1^2} = \frac{a^2 - c^2}{a^2} \tag{7.1.28}$$

it follows from Eq. (7.1.28) that

$$\bar{a} = \left[\frac{(a^2 - c^2)^2 + (\gamma - 1) b_2^2 c^2}{b_2^2 c^2 + (a^2 - c^2)^2}\right] \frac{a\bar{c}}{(\gamma - 1) c}.$$

Consequently, the first-order generalized Riemann invariants may be written in terms of $\bar{u}$ and $\bar{a}$, namely,

$$R = \frac{\bar{u}_1}{2} + \left[\frac{b_2^2 c^2 + (a^2 - c^2)^2}{(a^2 - c^2)^2 + (\gamma - 1) b_2^2 c^2}\right] \bar{a},$$

$$S = -\frac{\bar{u}_1}{2} + \left[\frac{b_2^2 c^2 + (a^2 - c^2)^2}{(a^2 - c^2)^2 + (\gamma - 1) b_2^2 c^2}\right] \bar{a}.$$

Thus

$$\bar{u}_1 + \bar{a} = \left\{\frac{3 (a^2 - c^2)^2 + (\gamma + 1) b_2^2 c^2}{2 [b_2^2 c^2 + (a^2 - c^2)^2]}\right\} \times$$

$$\times R + \left\{\frac{(\gamma - 3) b_2^2 c^2 - (a^2 - c^2)^2}{2 [b_2^2 c^2 + (a^2 - c^2)^2]}\right\} S,$$

$$\bar{u}_1 - \bar{a} = \left\{\frac{(a^2 - c^2)^2 - (\gamma - 3) b_2^2 c^2}{2 [b_2^2 c^2 + (a^2 - c^2)^2]}\right\} \times$$

$$\times R - \left\{\frac{3 (a^2 - c^2)^2 + (\gamma + 1) b_2^2 c^2}{2 [b_2^2 c^2 + (a^2 - c^2)^2]}\right\} S.$$

Exactly as in the case of a purely transverse field, the entropy perturbation may be determined independently and directly, i.e., from Eq. (7.1.27) which shows that $\bar{s}$ is constant along the curves obtained by integrating $dx/dt = u_1$. The solution is

$$\bar{s} = \Gamma(\Psi_0)$$

where Γ is an arbitrary differentiable function and $d\Psi_0 = \varrho\,[dx - u_1 dt]$. Exactly as in Chapter 1, the function T_0, defined by

$$c^2 \bar{s}_x = c^2 \varrho \Gamma'(\Psi_0) = \gamma (\gamma - 1) c_v T_0$$

6*

is introduced, so that the non-isentropic perturbation of an initially uniform or simple wave flow may be determined by finding solutions of the following two first-order linear equations for the first-order generalized Riemann invariants

$$R_t + (u_1 + a) R_x + \left[\left\{ \frac{3 (a^2 - c^2)^2 + (\gamma + 1) b_2^2 c^2}{2 [b_2^2 c^2 + (a^2 - c^2)^2]} \right\} \times \right.$$

$$\times R + \left. \left\{ \frac{(\gamma - 3) b_2^2 c^2 - (a^2 - c^2)^2}{2 [b_2^2 c^2 + (a^2 - c^2)^2]} \right\} S \right] \alpha_x \qquad (7.1.29)$$

$$= - \frac{u_1 a \bar{A}_x}{2 A} + \frac{T_0}{2} \, ,$$

$$S_t + (u_1 - a) S_x + \left[\left\{ \frac{(a^2 - c^2)^2 - (\gamma - 3) b_2^2 c^2}{2 [b_2^2 c^2 + (a^2 - c^2)^2]} \right\} \times \right.$$

$$\times R - \left. \left\{ \frac{3 (a^2 - c^2)^2 + (\gamma + 1) b_2^2 c^2}{2 [b_2^2 c^2 + (a^2 - c^2)^2]} \right\} S \right] \beta_x \qquad (7.1.30)$$

$$= - \frac{u_1 a \bar{A}_x}{2 A} - \frac{T_0}{2} \, .$$

For an initially uniform flow, α and β are constants, and the general solution of Eqs. (7.1.29)—(7.1.30) is

$$R = F [x - (u_1 + a) t] + \frac{E [x - u_1 t]}{a} - \frac{u_1 a \bar{A} (x)}{2 A (u_1 + a)} \, ,$$

$$S = G [x - (u_1 - a) t] + \frac{E [x - u_1 t]}{a} - \frac{u_1 a \bar{A} (x)}{2 A (u_1 - a)} \, ,$$

$$T_0 = 2 E' [x - u_1 t]$$

in terms of three arbitrary functions of one argument. This solution is directly analogous to the solution given by Eqs. (1.2.30)—(1.2.32), which it includes as a special case, and may be used to extend the analysis of Chapters 2, 3 and 4 to the oblique field case. Although the extensions will be given in a future publication, the next section is devoted to a relatively simple formulation of the transition relations across oblique shock waves. The final two sections of the chapter are concerned with simple wave perturbations.

7.2 Transition Relations Across Oblique Shock Waves

Transition relations across hydromagnetic shock waves have been considered by de HOFFMAN and TELLER [23], LÜST [24], HELFER [25], FRIEDRICHS [1], BAZER and ERICSON [26], [53], KANWAL [27], SHERCLIFF [55], GUNDERSEN [28], [29], and others. While all flow parameters behind a conventional gas dynamic shock wave may be determined by use of the Rankine-Hugoniot conditions if the flow in front of and one quantity behind the shock are specified, two parameters are required for a normal hydromagnetic shock wave, e.g., one giving the shock strength and one giving a measure of the applied transverse magnetic field. As shown in section 2.2, all flow quantities behind the shock may be expressed in terms of these and the known flow in front of the shock.

In the present section, it is shown that all flow quantities behind an oblique hydromagnetic shock wave may be expressed in terms of the flow quantities in front of the shock and three parameters, e.g., the shock strength, one giving a measure of the applied field and one giving a measure of the obliqueness of the applied field with respect to the shock front.

Let the shock front be perpendicular to the x-axis of an (x, y) coordinate system, and, as in section 7.1, let the velocity vector and induction have only two components. Further, let the flow velocity relative to the shock front, whose motion is parallel to the x-axis, be $v = u_1 - U$, and let the regions in front of and behind the shock front be denoted by the subscripts one and two, respectively. Then the hydromagnetic analogs of the jump conditions across oblique shock fronts are [1]

$$B_{11} = B_{12},\tag{7.2.1}$$

$$\varrho_1 v_1 = \varrho_2 v_2,\tag{7.2.2}$$

$$P_1 + \varrho_1 v_1^2 + \frac{B_{21}^2}{2\mu} = P_2 + \varrho_2 v_2^2 + \frac{B_{22}^2}{2\mu},\tag{7.2.3}$$

$$\varrho_1 v_1 u_{21} - B_{11}\frac{B_{21}}{\mu} = \varrho_2 v_2 u_{22} - \frac{B_{12}B_{22}}{\mu},\tag{7.2.4}$$

$$v_1 B_{21} - B_{11} u_{21} = v_2 B_{22} - B_{12} u_{22},\tag{7.2.5}$$

$$\frac{v_1^2}{2} - \frac{u_{21}^2}{2} + \frac{\gamma P_1}{(\gamma-1)\varrho_1} + \frac{B_{21}^2}{\varrho_1\mu}$$
$$= \frac{v_2^2}{2} - \frac{u_{22}^2}{2} + \frac{\gamma P_2}{(\gamma-1)\varrho_2} + \frac{B_{22}^2}{\varrho_2\mu}.\tag{7.2.6}$$

For the case of no applied field, Eqs. (7.2.1)—(7.2.6) reduce exactly to the transition relations given by COURANT and FRIEDRICHS [43], p. 299.

Introduce the parameters $\sigma = \varrho_2/\varrho_1$, $\tau = P_2/P_1$, $m_i = b_i/c$, $n_i = u_i/c$ and $B_{2i}/B_{1i} = \phi_i$, $i = 1, 2$. From Eq. (7.2.2), it follows that

$$\frac{\varrho_2}{\varrho_1} = \frac{v_1}{v_2} = \sigma.\tag{7.2.7}$$

In order to simplify the derivation, a coordinate system is chosen such that the shock front is at rest and the flow velocity and magnetic field are parallel on both sides of the shock front [1], [23]. This may be accomplished by choosing a coordinate system such that $\vec{u} = v\vec{B}/B_1$. Then Eq. (7.2.5) is satisfied identically and, further, there is no distinction between u_1 and v. Thus, it follows that

$$\frac{u_2}{u_1} = \frac{n_2}{n_1} = \frac{b_2}{b_1} = \frac{m_2}{m_1} = \frac{B_2}{B_1} = \phi\tag{7.2.8}$$

and

$$\frac{c_2^2}{c_1^2} = \frac{\tau}{\sigma} . \qquad (7.2.9)$$

From the continuity of the normal component of the magnetic field, i.e., Eq. (7.2.1) and Eqs. (7.2.7) and (7.2.8), the following relations are derived

$$\frac{B_{22}}{B_{21}} = \frac{\phi_2}{\phi_1} ,$$

$$\frac{b_{12}^2}{b_{11}^2} = \frac{1}{\sigma} ,$$

$$\frac{m_{22}^2}{m_{21}^2} = \frac{\phi_2^2}{\tau \phi_1^2} ,$$

$$\frac{b_{22}^2}{b_{21}^2} = \frac{\phi_2^2}{\sigma \phi_1^2} , \qquad (7.2.10)$$

$$\frac{m_{12}^2}{m_{11}^2} = \frac{1}{\tau} ,$$

$$\frac{n_{12}^2}{n_{11}^2} = \frac{1}{\sigma \tau} ,$$

$$\frac{n_{22}^2}{n_{21}^2} = \frac{\phi_2^2}{\sigma \tau \phi_1^2} .$$

Since $u_1 = v$ in the coordinate system utilized, Eq. (7.2.4) may be written as

$$\varrho_1 \phi_1 [v_1^2 - b_{11}^2] = \varrho_2 \phi_2 [v_2^2 - b_{12}^2]$$

or, solving for ϕ_2

$$\phi_2 = \frac{\sigma \phi_1 (n_{11}^2 - m_{11}^2)}{(n_{11}^2 - \sigma m_{11}^2)} \qquad (7.2.11)$$

which expresses ϕ_2 in terms of σ, ϕ_1, $m_{21} = \phi_1 m_{11}$ and the known flow in front of the shock.

Since Eqs. (7.2.1), (7.2.2), 7.2.4) and (7.2.5) have been utilized, only Eqs. (7.2.3) and (7.2.6) remain to be considered. These may be rewritten as

$$P_1 + \varrho_1 v_1^2 + \varrho_1 \frac{b_{21}^2}{2} = P_2 + \varrho_2 v_2^2 + \varrho_2 \frac{b_{22}^2}{2} ,$$

$$\frac{v_1^2 (1 - \phi_1^2)}{2} + \frac{\gamma P_1}{(\gamma - 1) \varrho_1} + b_{21}^2$$

$$= \frac{v_2^2 (1 - \phi_2^2)}{2} + \frac{\gamma P_2}{(\gamma - 1) \varrho_2} + b_{22}^2$$

or, if all terms with subscript two are expressed in terms of those with subscript one

$$\varrho_1 v_1^2 \left(1 - \frac{1}{\sigma} \right) + P_1 \left[1 - \tau + \frac{\gamma m_{21}^2}{2} \left(1 - \frac{\phi_2^2}{\phi_1^2} \right) \right] = 0 , \qquad (7.2.12)$$

$$\varrho_1 v_1^2 \left[1 - \phi_1^2 + \frac{(\phi_2^2 - 1)}{\sigma^2} \right] +$$

$$+ P_1 \left[\frac{2\gamma}{\gamma - 1} \left(1 - \frac{\tau}{\sigma} \right) + 2\gamma m_{21}^2 \left(1 - \frac{\phi_2^2}{\sigma \phi_1^2} \right) \right] = 0 . \qquad (7.2.13)$$

Since there exists a nontrivial solution to this linear system of homogeneous algebraic equations for P_1 and $\varrho_1 v_1^2$, Eqs. (7.2.12)—(7.2.13), the determinant of the coefficients must vanish, which leads to the result

$$\tau = \frac{[\sigma^2(1-\phi_1^2)+\phi_2^2-1]\left[1+\dfrac{\gamma m_{21}^2}{2}\left(1-\dfrac{\phi_2^2}{\phi_1^2}\right)\right]-2\gamma\sigma(\sigma-1)\left[\dfrac{1}{\gamma-1}+m_{21}^2\left(1-\dfrac{\phi_2^2}{\sigma\phi_1^2}\right)\right]}{\sigma^2(1-\phi_1^2)+\phi_2^2-1-\dfrac{2\gamma}{\gamma-1}(\sigma-1)} \qquad (7.2.14)$$

which expresses τ in terms of ϕ_1 and m_{21}. Eq. (7.2.14) gives the Hugoniot function for the case of an oblique magnetic field.

Because of Eqs. (7.2.11) and (7.2.13), Eqs. (7.2.7)—(7.2.10) express the flow quantities behind the shock in terms of those in front and the three parameters σ, ϕ_1 and m_{11}. In the limit of vanishing B_1, these results reduce to those presented in section 2.2 for the case of a normal shock. Some care must be exercised in carrying out this limit, e.g., it must be noted that

$$\lim_{B_1 \to 0} \frac{\phi_2}{\phi_1} = \sigma \ .$$

7.3 Perturbation of a Centered Simple Wave Flow

For a simple wave flow, one of the characteristic parameters of the base flow is constant throughout the flow. Let the wave be centered at the origin and be characterized by $\beta = \beta_0$, a constant. On each characteristic $dx/dt = u_1 + a$, the flow parameters are constant, and these characteristics are straight lines in the (x, t) plane. Consequently, the wave may be represented as

$$x = (u_1 + a)t \ , \qquad (7.3.1)$$

$$\beta = \beta_0 \ . \qquad (7.3.2)$$

From the definitions of the characteristic parameters of the base flow, it follows that

$$u_1 = \alpha - \beta_0 \ , \qquad (7.3.3)$$

$$w = \frac{(\gamma-1)}{2}(\alpha+\beta_0) \ . \qquad (7.3.4)$$

In Eq. (7.1.29), an expression for α_x is needed, and this may be obtained by differentiating Eq. (7.3.1) with respect to x which gives the result

$$\frac{1}{t} = u_{1x} + a_x = u_{1x} + \left[\frac{(a^2-c^2)^2 + (\gamma-1)b_2^2 c^2}{b_2^2 c^2 + (a^2-c^2)^2}\right]\frac{w_x}{\gamma-1} \ .$$

However, the derivatives u_{1x} and w_x may be expressed in terms of α_x through Eqs. (7.3.3)—(7.3.4). This yields the result

$$\alpha_x = \left[\frac{2\,[b_2^2 c^2 + (a^2 - c^2)^2]}{3\,(a^2 - c^2)^2 + (\gamma + 1)\,b_2^2 c^2} \right] t \,.$$ (7.3.5)

Consequently, for the wave centered at the origin, Eqs. (7.1.29)—(7.1.30) reduce to

$$R_t + (u_1 + a)\,R_x +$$

$$+ \left[R + \left\{ \frac{(\gamma - 3)\,b_2^2 c^2 - (a^2 - c^2)^2}{3\,(a^2 - c^2)^2 + (\gamma + 1)\,b_2^2 c^2} \right\} S \right] \frac{1}{t} = \frac{T_0}{2}\,,$$ (7.3.6)

$$S_t + (u_1 - a)\,S_x = -\frac{T_0}{2}$$ (7.3.7)

where attention will be restricted to simple wave flow in a uniform duct. Eqs. (7.3.6)—(7.3.7) will be solved by first considering the associated homogeneous system with $T_0 = 0$, and then the solution of the complete system may be obtained by adding a particular solution of the complete system.

The characteristics of the homogeneous equation associated with Eq. (7.3.7) are $dS = 0$ and

$$\frac{dx}{dt} = u_1 - a \,.$$ (7.3.8)

By the use of Eq. (7.3.1), Eq. (7.3.8) may be written as

$$\frac{dx}{dt} = u_1 + a + t\,\frac{d}{dt}\,(u_1 + a) = u_1 - a \,.$$ (7.3.9)

From the definition of the characteristic parameters

$$u_1 + a = a + \frac{2}{\gamma - 1} \int \frac{a}{c}\,dc - 2\beta_0$$

so that Eq. (7.3.9) may be written as

$$\frac{2\,dt}{t} + \frac{da}{a} + \frac{2}{\gamma - 1}\,\frac{dc}{c} = 0$$

which has the solution

$$a t^2 c^{\frac{2}{\gamma - 1}} = \text{constant}$$ (7.3.10)

or

$$\varrho a t^2 = \text{constant}\,.$$ (7.3.11)

Eq. (7.3.11) gives a convenient representation of the curvilinear cross-characteristics of the wave. In terms of an arbitrary differentiable function F, a convenient form for the solution of the homogeneous equation associated with Eq. (7.3.7) is

$$S = 2\,(\varrho a)^{\frac{1}{2}}\,t\,F'\,[\varrho a t^2]\,.$$ (7.3.12)

By the use of Eq. (7.3.1), the homogeneous equation associated with Eq. (7.3.6) may be written as

$$t R_t + x R_x + \left[R + \left\{ \frac{(\gamma - 3) b_2^2 c^2 - (a^2 - c^2)^2}{3 (a^2 - c^2)^2 + (\gamma + 1) b_2^2 c^2} \right\} S \right] = 0 . \qquad (7.3.13)$$

The solution for R may be obtained from the system

$$\frac{dt}{t} = \frac{dx}{x} = \frac{dR}{-R + \left[\dfrac{(a^2 - c^2)^2 - (\gamma - 3) b_2^2 c^2}{3 (a^2 - c^2)^2 + (\gamma + 1) b_2^2 c^2} \right] S} . \qquad (7.3.14)$$

From the first two ratios of Eq. (7.3.14), one first integral is

$$x/t = \text{constant} , \qquad (7.3.15)$$

i.e., the rectilinear characteristics. From the first and last ratios of Eq. (7.3.14), the following differential equation is obtained

$$\frac{d (t R)}{dt} = \left[\frac{(3 - \gamma) b_2^2 c^2 + (a^2 - c^2)^2}{3 (a^2 - c^2)^2 + (\gamma + 1) b_2^2 c^2} \right] \frac{d}{dt} \left[\frac{F [\varrho a t^2]}{(\varrho a)^{\frac{1}{2}}} \right] . \qquad (7.3.16)$$

In order to integrate Eq. (7.3.16), the previous first integral, Eq. (7.3.15), may be utilized, i.e., on $x/t = $ constant, the flow parameters are constant so that the solution of Eq. (7.3.16) may be written as

$$R t - \left[\frac{(3 - \gamma) b_2^2 c^2 + (a^2 - c^2)^2}{3 (a^2 - c^2)^2 + (\gamma + 1) b_2^2 c^2} \right] \frac{F [\varrho a t^2]}{(\varrho a)^{\frac{1}{2}}} = \text{constant} . \qquad (7.3.17)$$

From the first integrals, Eqs. (7.3.15) and (7.3.17), the solution of the homogeneous equation associated with Eq. (7.3.6) may be written as

$$R t = \left[\frac{(3 - \gamma) b_2^2 c^2 + (a^2 - c^2)^2}{3 (a^2 - c^2)^2 + (\gamma + 1) b_2^2 c^2} \right] \frac{F [\varrho a t^2]}{(\varrho a)^{\frac{1}{2}}} + G \left(\frac{x}{t} \right) \qquad (7.3.18)$$

where G is an arbitrary function.

Since the general solution to the homogeneous system associated with Eqs. (7.3.6)—(7.3.7) has been determined, the case of non-isentropic perturbed flow is reduced to finding a particular solution of the complete system, and this may be accomplished quite readily by choosing a convenient form for the function T_0.

The curves $\Psi_0 = $ constant are given by $\varrho a t = $ constant or, more conveniently, by

$$y = c^{\frac{1}{\gamma - 1}} a^{\frac{1}{2}} t^{\frac{1}{2}} = \text{constant} . \qquad (7.3.19)$$

Since

$$T_0 = \frac{\varrho c^2 \Gamma' (\Psi_0)}{\gamma (\gamma - 1) c_v} = \left[\frac{\varrho a t}{\gamma (\gamma - 1) c_v} \right] \frac{c^2}{a t} \Gamma' (\Psi_0)$$

it is convenient to write

$$T_0 = \frac{2 c^2 \Omega' (y)}{a t}$$

where Ω is an arbitrary differentiable function, and y is given by Eq. (7.3.19). Then the particular solution for S may be obtained from the equation

$$S_t + (u_1 - a) S_x = - \frac{c^2 \Omega'(y)}{at}$$

i.e., from the ratios

$$\frac{dt}{1} = \frac{dx}{u_1 - a} = \frac{dS}{-c^2 \Omega'(y)/at} . \tag{7.3.20}$$

From the two first ratios of Eq. (7.3.20), one first integral is

$$a^{\frac{1}{4}} t^{\frac{1}{2}} c^{1/2\,(\gamma-1)} = \lambda \tag{7.3.21}$$

where λ is a constant. From the first and last ratios of Eq. (7.3.20)

$$\frac{dS}{dt} = - \frac{c^2 \Omega'[a^{\frac{1}{2}} t^{\frac{1}{2}} c^{1/(\gamma-1)}]}{at} . \tag{7.3.22}$$

Using the first integral, Eq. (7.3.21), and the definition of y, Eq. (7.3.22) may be written as

$$\frac{dS}{dt} = \frac{2c^{\frac{2\gamma}{\gamma-1}}}{\lambda^6} t^{5/2} \frac{d\Omega[\lambda^2/t^{\frac{1}{2}}]}{dt} . \tag{7.3.23}$$

Integrating Eq. (7.3.23) and performing an integration by parts gives

$$S = \frac{2}{\lambda^6} c^{\frac{2\gamma}{\gamma-1}} t^{5/2} \Omega[\lambda^2/t^{\frac{1}{2}}] -$$
$$- \frac{2}{\lambda^6} \left\{ \int \Omega[\lambda^2/t^{\frac{1}{2}}] \, d\left[c^{\frac{2\gamma}{\gamma-1}} t^{5/2} \right] \right\}_\lambda + \text{constant} . \tag{7.3.24}$$

From the two first integrals, Eqs. (7.3.21) and (7.3.24), the particular solution for S may be written as

$$S = \frac{2c^2 \Omega(y)}{ay} - \frac{2}{t^{5/2} a y c^{\frac{2}{\gamma-1}}} \left\{ \int \Omega[\lambda^2/t^{\frac{1}{2}}] \, d\left[c^{\frac{2\gamma}{\gamma-1}} t^{5/2} \right] \right\}_\lambda =$$
$$= \frac{2c^2 \Omega(y)}{ay} + \Omega_1(x, t) \tag{7.3.25}$$

where the notation $\{\ \}_\lambda$ denotes that the quadrature is to be carried out with constant λ which must then be replaced by its equivalent from Eq. (7.3.21) to give the particular solution for S, and the quadrature is abbreviated by $\Omega_1(x, t)$.

The particular solution for R may be obtained from the equation

$$tR_t + xR_x + \left[R + \left\{ \frac{(\gamma-3)\,b_2^2 c^2 - (a^2-c^2)^2}{3\,(a^2-c^2)^2 + (\gamma+1)\,b_2^2 c^2} \right\} S \right] = \frac{c^2 \Omega'(y)}{a}$$

i.e., from the ratios

$$\frac{dt}{t} = \frac{dx}{x} = \frac{dR}{-R - \left\{\dfrac{(\gamma - 3)\, b_2^2 c^2 - (a^2 - c^2)^2}{3\,(a^2 - c^2)^2 + (\gamma + 1)\, b_2^2 c^2}\right\} S + \dfrac{c^2 \Omega'(y)}{a}} .$$

From the first two ratios, one first integral is

$$x/t = h \tag{7.3.26}$$

where h is constant. From the first and last ratios, the following differential equation is obtained

$$\frac{d\,[tR]}{dt} = \left[\frac{(3 - \gamma)\, b_2^2 c^2 + (a^2 - c^2)^2}{3\,(a^2 - c^2)^2 + (\gamma + 1)\, b_2^2 c^2}\right] \times$$

$$\times \left[\frac{2c^2 \Omega(y)}{ay} + \Omega_1(x, t)\right] + \frac{2c^2 t^{\frac{1}{2}}}{a^{3/2} c^{\frac{1}{\gamma - 1}}} \frac{d\Omega}{dt} . \tag{7.3.27}$$

To integrate Eq. (7.3.27), the first integral, Eq. (7.3.26) may be utilized, i.e., the flow parameters are constant on the rectilinear characteristics. This gives

$$Rt = \left[\frac{(3 - \gamma)\, b_2^2 c^2 + (a^2 - c^2)^2}{3\,(a^2 - c^2)^2 + (\gamma + 1)\, b_2^2 c^2}\right] \int \Omega_1(ht, t)\, dt +$$

$$+ \left[\frac{(3 - \gamma)\, b_2^2 c^2 + (a^2 - c^2)^2}{3\,(a^2 - c^2)^2 + (\gamma + 1)\, b_2^2 c^2}\right] \frac{2c^2}{a^{3/2} c^{1/(\gamma - 1)}} \int \frac{\Omega(y)\, dt}{t^{\frac{1}{2}}} + \tag{7.3.28}$$

$$+ \frac{2c^2}{a^{3/2} c^{1/(\gamma - 1)}} \int t^{\frac{1}{2}} \frac{d\Omega(y)}{dt}\, dt + \text{constant} .$$

Performing an integration by parts in the third term on the right-hand side of Eq. (7.3.28) and using this result and the first integral, Eq. (7.3.26), the final particular solution for R may be written as

$$Rt = \frac{2c^2 t \Omega(y)}{ay} + \frac{c^2 t^{\frac{1}{2}}}{ay}\left[\frac{(5 - 3\gamma)\, b_2^2 c^2 - (a^2 - c^2)^2}{3\,(a^2 - c^2)^2 + (\gamma + 1)\, b_2^2 c^2}\right] \times$$

$$\times \left\{\int \frac{\Omega(y)\, dt}{t^{\frac{1}{2}}}\right\}_{h = x/t} + \left[\frac{(3 - \gamma)\, b_2^2 c^2 + (a^2 - c^2)^2}{3\,(a^2 - c^2)^2 + (\gamma + 1)\, b_2^2 c^2}\right] \times \tag{7.3.29}$$

$$\times \left\{\int \Omega_1(ht, t)\, dt\right\}_{h = x/t}$$

where the notation $\{\ \}_{h = x/t}$ denotes that the quadrature is to be carried out with x/t constant; then, replacing h by x/t, the solution, Eq. (7.3.29), is obtained.

The general solution for the non-isentropic perturbation of the centered simple wave is given by Eq. (7.3.12) plus Eq. (7.3.25) and Eq. (7.3.18) plus Eq. (7.3.29).

7.4 Perturbation of an Arbitrary Simple Wave

Suppose the wave is characterized by $\beta = \beta_0$, a constant. Then letting $[x_0(z), t_0(z)]$ be the parametric representation of a curvilinear characteristic of the simple wave, the wave may be represented by

$$x = x_0(z) + [u_1(z) + a(z)]\tau, \quad t = t_0(z) + \tau, \tag{7.4.1}$$

$$-\frac{u_1(z)}{2} + \frac{w(z)}{\gamma - 1} = \beta_0. \tag{7.4.2}$$

The pertinent perturbation equations are Eqs. (7.1.29)—(7.1.30), and because of Eq. (7.4.2), Eq. (7.1.30) reduces to

$$S_t + (u_1 - a)S_x = -\frac{T_0}{2}. \tag{7.4.3}$$

In order to solve Eq. (7.4.3), it is convenient to change from the independent variables (x, t) to (z, τ). The transformation is effectively the same as carried out in section (5.4), and, with the definition that $S(x, t) = S_1(z, \tau)$, it follows that

$$J S_x = S_{1z} - t_0' S_{1\tau},$$

$$J S_t = [x_0' + \tau(u_1' + a')] S_{1\tau} - (u_1 + a) S_{1z},$$

$$J = -2a t_0' + \tau(u_1' + a').$$

From the definition of the characteristic parameters,

$$u_1 + a = a + \frac{2}{\gamma - 1} \int \frac{a}{c} dc - 2\beta_0$$

so that

$$u_1' + a' = \left[\frac{3(a^2 - c^2)^2 + (\gamma + 1) b_2^2 c^2}{b_2^2 c^2 + (a^2 - c^2)^2}\right] \frac{ac'}{(\gamma - 1)c} \tag{7.4.4}$$

and the Jacobian may be written as

$$J = -2a t_0' + \left[\frac{3(a^2 - c^2)^2 + (\gamma + 1) b_2^2 c^2}{b_2^2 c^2 + (a^2 - c^2)^2}\right] \frac{ac'\tau}{(\gamma - 1)c}. \tag{7.4.5}$$

Under the prescribed transformation of independent variables, Eq.(7.4.3) becomes

$$-2a S_{1z} + \left[\frac{3(a^2 - c^2)^2 + (\gamma + 1) b_2^2 c^2}{b_2^2 c^2 + (a^2 - c^2)^2}\right] \times$$
$$\times \frac{ac'\tau}{(\gamma - 1)c} S_{1\tau} = -\frac{J T_0}{2}. \tag{7.4.6}$$

Since the curves $\Psi_0 = $ constant are given by $\varrho a\tau = $ constant or, more conveniently, by

$$y = c^{\frac{1}{\gamma - 1}} a^{\frac{1}{2}} \tau^{\frac{1}{2}} = \text{constant}, \tag{7.4.7}$$

it is convenient to write

$$\frac{T_0}{2} = \frac{c^2 \Omega'(y)}{a\tau} \tag{7.4.8}$$

where Ω is an arbitrary differentiable function. Consequently, the solution of Eq. (7.4.6) may be obtained from the ratios

$$\frac{dz}{-2a} = \frac{d\tau}{\left[\dfrac{3(a^2 - c^2)^2 + (\gamma + 1) b_2^2 c^2}{b_2^2 c^2 + (a^2 - c^2)^2}\right]\dfrac{a\tau c'}{(\gamma - 1)c}} = \frac{dS_1}{-\dfrac{J T_0}{2}}. \tag{7.4.9}$$

From the first two ratios of Eq. (7.4.9), the following first integral is obtained

$$c^{\frac{1}{2(\gamma-1)}} a^{\frac{1}{4}} \tau^{\frac{1}{2}} = \lambda \tag{7.4.10}$$

where λ is a constant or

$$\varrho a \tau^2 = \text{constant} . \tag{7.4.11}$$

Eq. (7.4.11) gives a convenient representation of the curvilinear cross-characteristics of the simple wave. From the first and third ratios of Eq. (7.4.9), the following differential equation is obtained

$$2 d S_1 = \frac{c \Omega'(y)}{(\gamma - 1) a} \left[\frac{3(a^2 - c^2)^2 + (\gamma + 1) b_2^2 c^2}{b_2^2 c^2 + (a^2 - c^2)^2}\right] dc - \frac{2c^2 \Omega'(y) dt_0}{a\tau} . \tag{7.4.12}$$

In order to integrate Eq. (7.4.12), the previous first integral may be used, and this gives a relation between z and τ, namely,

$$\frac{dz}{d\tau} = -\frac{2(\gamma - 1)c[b_2^2 c^2 + (a^2 - c^2)^2]}{\tau[3(a^2 - c^2)^2 + (\gamma + 1) b_2^2 c^2]c'} \tag{7.4.13}$$

and from the definition of y

$$y = \lambda^2/\tau^{\frac{1}{2}} .$$

Consequently, Eq. (7.4.12) may be written as

$$\frac{\lambda^6 d S_1}{2} = \tau^{5/2} c^{\frac{2\gamma}{\gamma-1}} \frac{d\Omega\left[\lambda^2/\tau^{\frac{1}{2}}\right]}{d\tau} dt_0 + \tau^{5/2} c^{\frac{2\gamma}{\gamma-1}} d\Omega . \tag{7.4.14}$$

Integrating Eq. (7.4.14), performing an integration by parts in the first term and introducing Eq. (7.4.13) in the second gives

$$\frac{\lambda^6}{2} S_1 = c^{\frac{2\gamma}{\gamma-1}} \tau^{\frac{5}{2}} \Omega\left[\lambda^2/\tau^{\frac{1}{2}}\right] - \int \Omega\left[\lambda^2/\tau^{\frac{1}{2}}\right] d\left[c^{\frac{2\gamma}{\gamma-1}} \tau^{\frac{5}{2}}\right] -$$

$$- \int \frac{c^{\frac{2\gamma}{\gamma-1}} \tau^{\frac{5}{2}} 2(\gamma - 1) c t_0'[b_2^2 c^2 + (a^2 - c^2)^2] d\Omega}{\tau[3(a^2 - c^2)^2 + (\gamma + 1) b_2^2 c^2]c'} + \text{constant} .$$

Performing an integration by parts in the last integral gives the following first integral

$$S_1 = \frac{2}{\lambda^6} c^{\frac{2\gamma}{\gamma-1}} \tau^{\frac{5}{2}} \Omega\left[\lambda^2/\tau^{\frac{1}{2}}\right] - \frac{2}{\lambda^6}\left\{\left\{\int\int \Omega\left[\lambda^2/\tau^{\frac{1}{2}}\right] d\left[c^{\frac{2\gamma}{\gamma-1}}\tau^{\frac{5}{2}}\right]\right\}_\lambda -$$

$$- \frac{4(\gamma-1)}{\lambda^6} c^{\frac{3\gamma-1}{\gamma-1}} \tau^{\frac{3}{2}} \Omega\left[\lambda^2/\tau^{\frac{1}{2}}\right] \left[\frac{b_2^2 c^2 + (a^2-c^2)^2}{3(a^2-c^2)^2 + (\gamma+1) b_2^2 c^2}\right] \times$$

$$\times \frac{t_0'}{c'} + \frac{4(\gamma-1)}{\lambda^6}\left\{\left\{\int\int \Omega\left[\lambda^2/\tau^{\frac{1}{2}}\right] d\times\right.\right.$$

$$\left.\left. \times \left[c^{\frac{3\gamma-1}{\gamma-1}} \tau^{\frac{3}{2}} \frac{t_0'}{c'} \left\{\frac{b_2^2 c^2 + (a^2-c^2)^2}{3(a^2-c^2)^2 + (\gamma+1) b_2^2 c^2}\right\}\right]\right\}_\lambda + \text{constant}.$$

(7.4.15)

From the two first integrals, Eqs. (7.4.11) and (7.4.15), the solution of Eq. (7.4.3) may be written as

$$S_1 = 2(\varrho a)^{\frac{1}{3}} \tau F'[\varrho a\tau^2] + \frac{2c^2\Omega(y)}{ay\tau}\left[\tau - 2(\gamma-1) c\frac{t_0'}{c'} \times\right.$$

$$\left. \times \left\{\frac{b_2^2 c^2 + (a^2-c^2)^2}{3(a^2-c^2)^2 + (\gamma+1) b_2^2 c^2}\right\}\right] - \frac{2}{c^{\frac{2}{\gamma-1}} \tau^{\frac{5}{2}} ay} \times$$

$$\times \left\{\left\{\int\int \Omega\left[\lambda^2/\tau^{\frac{1}{2}}\right] d\left[c^{\frac{2\gamma}{\gamma-1}} \tau^{\frac{3}{2}} \left\{\tau - 2(\gamma-1) c\frac{t_0'}{c'} \times\right.\right.\right.\right.$$ (7.4.16)

$$\left.\left.\left.\left. \times \frac{[b_2^2 c^2 + (a^2-c^2)^2]}{3(a^2-c^2)^2 + (\gamma+1) b_2^2 c^2}\right\}\right]\right\}\right\}_\lambda \equiv 2(\varrho a)^{\frac{1}{3}} \tau F'[\varrho a\tau^2] +$$

$$+ \frac{2c^2\Omega(y)}{ay\tau}\left[\tau - 2(\gamma-1) \frac{ct_0'}{c'}\left\{\frac{b_2^2 c^2 + (a^2-c^2)^2}{3(a^2-c^2)^2 + (\gamma+1) b_2^2 c^2}\right\}\right] + \Phi(z,\tau)$$

where F is an arbitrary differentiable function, $\{\ \}_\lambda$ has the same significance as in Eq. (7.3.25) and Φ is used to denote the quadrature.

From the definition of the characteristic parameters of the base flow and Eq. (7.4.4), it follows that

$$\alpha_x = \frac{2ac'}{(\gamma-1)cJ} = \frac{\dfrac{2[b_2^2 c^2 + (a^2-c^2)^2]}{3(a^2-c^2)^2 + (\gamma+1) b_2^2 c^2}}{\tau - \dfrac{2(\gamma-1) ct_0'[b_2^2 c^2 + (a^2-c^2)^2]}{c'[3(a^2-c^2)^2 + (\gamma+1) b_2^2 c^2]}}.$$ (7.4.17)

Substitution of Eq. (7.4.17) into Eq. (7.1.29) gives the equation for R, viz.,

$$R_t + (u_1 + a) R_x +$$

$$+ \left[R + S\left\{\frac{(\gamma-3) b_2^2 c^2 - (a^2-c^2)^2}{3(a^2-c^2)^2 + (\gamma+1) b_2^2 c^2}\right\}\right]\frac{1}{\nu} = \frac{T_0}{2}$$ (7.4.18)

where $\nu \equiv \tau - \dfrac{2(\gamma-1) ct_0'[b_2^2 c^2 + (a^2-c^2)^2]}{c'[3(a^2-c^2)^2 + (\gamma+1) b_2^2 c^2]}$.

The solution of Eq. (7.4.18) may be obtained from the ratios

$$\frac{dt}{1} = \frac{dx}{u_1 + a}$$

$$= \frac{dR}{-\dfrac{R}{\nu} - \dfrac{[(\gamma - 3)\, b_2^2 c^2 - (a^2 - c^2)^2]\, S}{[3\,(a^2 - c^2)^2 + (\gamma + 1)\, b_2^2 c^2]\, \nu} + \dfrac{T_0}{2}} \; . \tag{7.4.19}$$

The integral of the first two ratios of Eq. (7.4.19) has the rectilinear characteristics for level curves. This first integral may be written as

$$z(x, t) = h \tag{7.4.20}$$

where h is a constant, and the function is defined implicitly by

$$x - x_0(z) = [u_1(z) + a(z)]\,[t - t_0(z)]\,.$$

From the first and last ratios of Eq. (7.4.19), the following differential equation is obtained

$$\frac{d}{dt}\,[\nu R] = \left[\frac{(a^2 - c^2)^2 - (\gamma - 3)\, b_2^2 c^2}{3\,(a^2 - c^2)^2 + (\gamma + 1)\, b_2^2 c^2}\right] \frac{d}{dt}\left[\frac{F(\varrho a \tau^2)}{(\varrho a)^{\frac{1}{2}}}\right] +$$

$$+ \left[\frac{(a^2 - c^2)^2 - (\gamma - 3)\, b_2^2 c^2}{3\,(a^2 - c^2)^2 + (\gamma + 1)\, b_2^2 c^2}\right] \Phi(h, t - t_0) + \tag{7.4.21}$$

$$+ \left[\frac{(a^2 - c^2)^2 - (\gamma - 3)\, b_2^2 c^2}{3\,(a^2 - c^2)^2 + (\gamma + 1)\, b_2^2 c^2}\right] \frac{2 c^2 \nu \,\Omega(y)}{a y\,(t - t_0)} + \frac{c^2 \nu\, \Omega'(y)}{a\,(t - t_0)}\,.$$

In order to integrate Eq. (7.4.21), the first integral Eq. (7.4.20) may be used, i.e., z may be treated as a constant. This gives

$$\nu R = \left[\frac{(a^2 - c^2)^2 - (\gamma - 3)\, b_2^2 c^2}{3\,(a^2 - c^2)^2 + (\gamma + 1)\, b_2^2 c^2}\right] \frac{F[\varrho a \tau^2]}{(\varrho a)^{\frac{1}{2}}} +$$

$$+ \left[\frac{(a^2 - c^2)^2 - (\gamma - 3)\, b_2^2 c^2}{3\,(a^2 - c^2)^2 + (\gamma + 1)\, b_2^2 c^2}\right] \frac{2 c^2}{a} \int \frac{\Omega(y)\, \nu\, dt}{y\,(t - t_0)} +$$

$$+ \left[\frac{(a^2 - c^2)^2 - (\gamma - 3)\, b_2^2 c^2}{3\,(a^2 - c^2)^2 + (\gamma + 1)\, b_2^2 c^2}\right] \int \Phi(h, t - t_0)\, dt + \tag{7.4.22}$$

$$+ \frac{c^2}{a} \int \frac{\Omega'(y)\, \nu\, dt}{t - t_0} + \text{constant}\,.$$

After an integration by parts is performed in the last integral in Eq. (7.4.22), the following solution is obtained from Eqs. (7.4.20) and (7.4.22)

$$\nu R = G\,[z] + \left[\frac{(a^2 - c^2)^2 - (\gamma - 3)\, b_2^2 c^2}{3\,(a^2 - c^2)^2 + (\gamma + 1)\, b_2^2 c^2}\right] \frac{F[\varrho a \tau^2]}{(\varrho a)^{\frac{1}{2}}} +$$

$$+ \frac{2 c^2 \nu\, \Omega(y)}{a y} + \left\{ \frac{2 c^2}{a} \left[\frac{(a^2 - c^2)^2 - (\gamma - 3)\, b_2^2 c^2}{3\,(a^2 - c^2)^2 + (\gamma + 1)\, b_2^2 c^2}\right] \times \right.$$

$$\times \int \frac{\Omega(y)\, \nu\, dt}{y\,(t - t_0)} + \left[\frac{(a^2 - c^2)^2 - (\gamma - 3)\, b_2^2 c^2}{3\,(a^2 - c^2)^2 + (\gamma + 1)\, b_2^2 c^2}\right] \times$$

$$\times \int \Phi(h, t - t_0)\, dt - \frac{2 c^2}{a} \int \Omega(y)\, d\,[\nu/y] \Bigg\}_{h = z}$$

where G is an arbitrary function and $\{\ \}_{h = z}$ has the same significance as in Eq. (7.3.25).

References

[1] FRIEDRICHS, K. O.: Nonlinear wave motion in magnetohydrodynamics. Los Alamos Report 2105, 1957. See also a later version of this report by K. O. FRIEDRICHS and H. KRANZER: Nonlinear wave motion. A. E. C. Research and Development Report, NYO-6486, 1958.

[2] GERMAIN, P., and R. GUNDERSEN: Sur les écoulements unidimensionnels d'un fluide parfait à entropie faiblement variable. C. R. Acad. Sci. (Paris) 241, 925 (1955).

[3] LUNDQUIST, S.: Studies in magnetohydrodynamics. Arkiv für Fysik 5, 297 (1952).

[4] GUNDERSEN, R.: The propagation of non-uniform magnetohydrodynamic shocks, with special reference to cylindrical and spherical shock waves. MRC-TS-310, 1962. Published in Arch. Rat. Mech. Anal. 11, 1 (1962).

[5] — The flow of a compressible fluid with weak entropy changes. Thesis, Brown Univ., Mar., 1956.

[6] — The flow of a compressible fluid with weak entropy changes. J. Fluid Mech. 3, 553 (1958).

[7] — Shock propagation in time-dependent ducts. J. Aerospace Sci. 28, 748 (1961).

[8] CHESTER, W.: The quasi-cylindrical shock tube. Phil. Mag. (7) 45, 1293 (1954).

[9] CHISNELL, R.: The motion of a shock wave in a channel, with applications to cylindrical and spherical shock waves. J. Fluid Mech. 2, 286 (1957).

[10] WHITHAM, G.: On the propagation of shock waves through regions of non-uniform area or flow. J. Fluid Mech. 4, 337 (1958).

[11] ROSCISZEWSKI, J.: Calculations of the motion of non-uniform shock waves. J. Fluid Mech. 8, 337 (1960).

[12] GUNDERSEN, R.: A note on shock flow in a channel. J. Fluid Mech. 4, 501 (1958).

[13] STOCKER, P.: The transients arising from the addition of heat to a gas flow. Proc. Camb. Phil. Soc. 48, 482 (1952).

[14] FRIEDMAN, M.: An improved perturbation theory for shock waves propagating through non-uniform regions. J. Fluid Mech. 8, 193 (1960).

[15] GUNDERSEN, R.: Secondary shocks in magnetohydrodynamic channel flow. Int. J. Eng. Sci. 1, 241 (1963).

[16] — The propagation of a plane shock wave into a moving fluid. J. Aerospace Sci. 28, 584 (1961).

[17] — A perturbation analysis of shock flow in a nozzle. J. Aerospace Sci. 26, 763 (1959).

[18] MIRELS, H.: Source distribution for unsteady one-dimensional flows with small mass, momentum, and heat addition and small area variation. NASA Memo 5-6-59E, 1959.

[19] GUNDERSEN, R.: The piston-driven shock. J. Aerospace Sci. 27, 467 (1960).

[20] — Magnetohydrodynamic shock propagation in non-uniform ducts. MRC-TS-287, 1961; Z.A.M.P., 14, 124 (1963).

[21] — Cylindrical and spherical shock waves in monatomic conducting fluids. MRC-TS-299, 1962. Appl. Sci. Res., 10 B, 119 (1963).

[22] — The propagation of a non-uniform magnetohydrodynamic shock wave into a moving monatomic fluid. J. Aerospace Sci. 29, 1421 (1962).

[23] DE HOFFMAN, F., and E. TELLER: Magnetohydrodynamic shocks. Phys. Rev. 80, 692 (1950).

[24] LÜST, R.: Magneto-hydrodynamische Stoßwellen in einem Plasma unendlicher Leitfähigkeit. Z. Naturforsch. 8a, 277 (1953).

[25] HELFER, L.: Magnetohydrodynamic shock waves. Astrophys. J. **117**, 177 (1953).

[26] BAZER, J., and W. ERICSON: Hydromagnetic shocks. Astrophys. J. **129**, 758 (1958).

[27] KANWAL, R.: On magnetohydrodynamic shock waves. J. Math. Mech. **9**, 681 (1960).

[28] GUNDERSEN, R.: Normal shock relations in magnetohydrodynamics. J. Aerospace Sci. **29**, 997 (1962).

[29] — Transition relations across oblique magnetohydrodynamic shock waves. A.I.A.A.J. **1**, 482 (1963).

[30] — The non-isentropic perturbation of a centered magnetohydrodynamic simple wave. MRC-TS-280, 1961. J. Math. Anal. Appl. **6**, 86 (1963).

[31] — Entropy perturbations in one-dimensional magnetohydrodynamic flow. A.I.A.A.J. **1**, 969 (1963).

[32] — Quasi-one-dimensional magnetohydrodynamic flow with heat addition. Z.A.M.P. **14**, 294 (1963).

[33] LIN, C. C.: Note on the characteristics in unsteady one-dimensional flows with heat addition. Q. A. M. **7**, 443 (1950).

[34] FOA, J., and G. RUDINGER: On the addition of heat to a gas flowing at subsonic speeds. J. Aero. Sci. **16**, 84 (1949).

[35] — — Heat addition to a flowing gas. J. Aero. Sci. **16**, 566 (1949).

[36] CHAMBRÉ, P., and C. C. LIN: On the steady flow of a gas through a tube with heat exchange or chemical reaction. J. Aero. Sci. **13**, 537 (1946).

[37] GUNDERSEN, R.: The effects of heat addition to one-dimensional magnetohydrodynamic flow. Int. J. Eng. Sci. **1**, 359 (1963).

[38] — The non-isentropic perturbation of an arbitrary simple wave. J. Math. Mech. **9**, 141 (1960).

[39] — Simple wave flow in ducts. Phys. Fluids **2**, 680 (1959).

[40] — Hydromagnetic simple wave flow in non-uniform ducts. J. Math. Anal. Appl. **6**, 277 (1963).

[41] ROSCISZEWSKI, J.: Propagation of waves of finite amplitude along a duct of non-uniform cross-section. J. Fluid. Mech. **8**, 625 (1960).

[42] FRIEDMAN, M.: A simplified analysis of spherical and cylindrical blast waves. J. Fluid Mech. **11**, 1 (1961).

[43] COURANT, R., and K. O. FRIEDRICHS: Supersonic flow and shock waves. New York: Interscience Publ. 1948.

[44] RIEMANN, B.: Über die Fortpflanzung ebener Luftwellen von endlicher Schwingungsweite. Gesammelte Mathematische Werke. Leipzig 1892.

[45] COHN, H.: The Rieman Function for $\dfrac{\partial^2 u}{\partial x \partial y} + H(x+y)u = 0$. Duke Math. J. **14**, 297 (1947).

[46] FRIEDRICHS, K.: Formation and decay of shock waves. Comm. Pure Appl. Math. **1**, 211 (1948).

[47] LIGHTHILL, M.: The energy distribution behind decaying shock waves. I. Plane waves. Phil. Mag. (7), **41**, 1101 (1950).

[48] MARTIN, M.: A new approach to problems in two dimensional flow. Quart. Appl. Math. **8**, 137 (1950).

[49] MISES, R. VON: Mathematical theory of compressible fluid flow. New York: Academic Press 1958.

[50] WEIR, D.: A family of exact solutions of one-dimensional anisentropic flow. Proc. Camb. Phil. Soc. **57**, 890 (1961).

[51] GUNDERSEN, R.: A class of exact solutions of non-isentropic one-dimensional magnetohydrodynamic flow. A.I.A.A.J. **1**, 1191 (1963).

[52] — The characteristic form of the equations of one-dimensional magneto-hydrodynamic flow with oblique magnetic field. A.I.A.A.J. **1**, 219 (1962).

[53] ERICSON, W., and J. BAZER: On certain properties of hydromagnetic shocks. Phys. Fluids **3**, 631 (1960).

[54] BAZER, J.: Resolution of an initial shear-flow discontinuity in one-dimensional hydromagnetic flow. Astrophys. J. **128**, 686 (1958).

[55] SHERCLIFF, J.: One-dimensional magnetohydrodynamics in oblique fields. J. Fluid Mech. **9**, 481 (1960).

[56] HERLOFSON, N.: Magnetohydrodynamic waves in a compressible fluid conductor. Nature (Lond.) **165**, 1020 (1950).

[57] VAN DE HULST, H.: Interstellar polarization and magnetohydrodynamic waves in problems of cosmical aerodynamics. Int. Union Theor. Appl. Mech. and Int. Astr. Union. Proc. Symposium. Paris 1949.

[58] GRAD, H.: Reducible problems in magneto-fluid dynamic steady flows. Rev. Mod. Phys. **32**, 830 (1960).

Appendix A

Principal Notation

x	$=$	distance
t	$=$	time
u	$=$	particle velocity
c	$=$	local speed of sound
s	$=$	specific entropy
s^*	$=$	specific entropy at some reference state
ϱ	$=$	density
$\vec{B} = (0,\, 0,\, B)$	$=$	magnetic induction
μ	$=$	permeability
$b^2 = \dfrac{B^2}{\mu\,\varrho}$	$=$	square of Alfvén speed
P	$=$	pressure
γ	$=$	adiabatic index, ratio of specific heat at constant pressure c_p and at constant volume c_v
A	$=$	cross-sectional area of the channel
$\omega = (b^2 + c^2)^{\frac{1}{2}}$	$=$	true speed of sound (the limiting case of a fast wave)
$(\alpha,\, \beta,\, \xi)$	$=$	characteristic parameters
η^*	$= \dfrac{2\,(2-\gamma)}{\gamma-1}$	
R	$= \dfrac{u_1}{2} + \dfrac{w_1}{\gamma-1}$	
S	$= -\dfrac{u_1}{2} + \dfrac{w_1}{\gamma-1}$	
E, F, G	$=$	arbitrary functions
σ	$= \dfrac{\varrho_2}{\varrho_1}$	
τ	$= \dfrac{P_2}{P_1}$	
U	$=$	shock velocity
v	$= U - u$	
m	$= \dfrac{b}{c}$	
n	$= \dfrac{u}{c}$	
M	$= \dfrac{v}{c}$	
q^2	$= \dfrac{\omega^2}{c^2} = 1 + m^2$	
θ	$= \dfrac{\gamma+1}{\gamma-1}\,.$	

Appendix B

The Characteristic Form of the Basic Equations

An outline of the Courant-Friedrichs method [43] for obtaining the characteristic form of the basic equations (1.2.2)—(1.2.5) for the case of no area variations will be given.

If Eqs. (1.2.2)—(1.2.5) are multiplied by λ_1, λ_2, λ_3 and λ_4, respectively, the resulting equations added, and it is required then that the derivatives of u, c, B and s enter only as directional derivatives with respect to ε, there results

$$\frac{2\lambda_1 c_t}{\gamma - 1} + \left[\frac{2\lambda_1 u}{\gamma - 1} + \frac{2\lambda_2 c}{\gamma - 1}\right] c_x = c_\varepsilon , \tag{B1}$$

$$\lambda_2 u_t + [\lambda_1 c + \lambda_2 u + \lambda_3 B] u_x = u_\varepsilon , \tag{B2}$$

$$\lambda_3 B_t + \left[\lambda_2 \frac{b^2}{B} + \lambda_3 u\right] B_x = B_\varepsilon , \tag{B3}$$

$$\lambda_4 s_t + \left[\lambda_4 u - \frac{\lambda_2 c^2}{\gamma (\gamma - 1) c_v}\right] s_x = s_\varepsilon . \tag{B4}$$

From (B1)—(B4)

$$\frac{x_\varepsilon}{t_\varepsilon} = \frac{\lambda_1 u + \lambda_2 c}{\lambda_1} = \frac{\lambda_1 c + \lambda_2 u + \lambda_3 B}{\lambda_2}$$
$$= \frac{\lambda_2 b^2/B + \lambda_3 u}{\lambda_3} = \frac{\lambda_4 u - \lambda_2 c^2/\gamma (\gamma - 1) c_v}{\lambda_4} . \tag{B5}$$

The Eqs. (B5) lead to the following three independent relations for non-vanishing λ_j $(j = 1, 2, 3, 4)$

$$c \lambda_2^2 = c \lambda_1^2 + \lambda_1 \lambda_3 B \tag{B6}$$

$$c \lambda_3 = \frac{\lambda_1 b^2}{B} , \tag{B7}$$

$$\lambda_4 = - \frac{\lambda_1 c}{\gamma (\gamma - 1) c_v} . \tag{B8}$$

If $\lambda_2 = 0$, then Eqs. (B5) lead to two relations when λ_4 is arbitrary but λ_1 and λ_3 do not vanish. Thus

$$\lambda_2 = 0 , \quad \lambda_1 c + \lambda_3 B = 0 . \tag{B9}$$

The cases when λ_1 and/or λ_3 and/or λ_4 vanish lead to no new equations. If the variable $\omega^2 = b^2 + c^2$ is introduced, then after eliminating $\frac{\lambda_3}{\lambda_1}$ in

(B6) by use of (B7), the following two relations are obtained

$$\lambda_2 = \frac{\lambda_1 \omega}{c} , \tag{B10}$$

$$\lambda_2 = -\frac{\lambda_1 \omega}{c} . \tag{B11}$$

Consequently, there are three basic types of relations for λ_j which follow from Eq. (B5): (1) the Eqs. (B7), (B8) and (B10); (2) the Eqs. (B7), (B8) and (B11); (3) the Eqs. (B9) with λ_4 arbitrary. From Eq. (B5), it is seen that if the curves of the above three families are parameterized by β = variable, α = variable, ξ = variable, respectively, then

$$x_\beta = (u + \omega) t_\beta , \tag{B12}$$

$$x_\alpha = (u - \omega) t_\alpha , \tag{B13}$$

$$x_\xi = u t_\xi . \tag{B14}$$

Thus Eqs. (B12) and (B13) are the C_+ and C_- characteristics and (B14) are the streamlines.

Now, returning to Eqs. (1.2.2)—(1.2.5) and multiplying them respectively by λ_1, λ_2, λ_3 and λ_4, adding them and using the relations (B7), (B8), (B10) and (B12) gives

$$(\omega^2 - c^2) \frac{B_\beta}{B} + \omega u_\beta + \frac{2 c c_\beta}{(\gamma - 1)} - \frac{c^2 s_\beta}{\gamma (\gamma - 1) c_v} = 0 . \tag{B15}$$

Similarly, from Eqs. (B7), (B8), (B11) and (B13), there results

$$(\omega^2 - c^2) \frac{B_\alpha}{B} - \omega u_\alpha + \frac{2 c c_\alpha}{\gamma - 1} - \frac{c^2 s_\alpha}{\gamma (\gamma - 1) c_v} = 0 . \tag{B16}$$

Finally, by use of Eqs. (B9) and (B14)

$$\frac{B_\xi}{B} - \frac{2 c_\xi}{(\gamma - 1) c} = 0 . \tag{B17}$$

The Eqs. (B15)—(B17) and

$$s_\xi = 0 \tag{B18}$$

are the characteristic forms of Eqs. (1.2.2)—(1.2.5).

Appendix C

Tables are presented for the parameters

$$K(\sigma, m_1, n_1),$$
$$Y(\sigma, m_1, n_1),$$
$$Z(\sigma, m_1, n_1)$$

for various values of the arguments.

Sigma	$K(\sigma, .0, .00)$	$Y(\sigma, .0, .00)$	$Z(\sigma, .0, .00)$
1.10	.483328	− .100901	.584229
1.20	.471484	− .206667	.678151
1.30	.462696	− .321260	.783955
1.40	.455940	− .448485	.904425
1.50	.450593	− .592593	1.043186
1.60	.446251	− .758824	1.205075
1.70	.442647	− .954074	1.396721
1.80	.439593	− 1.187879	1.627472
1.90	.436960	− 1.474016	1.910976
2.00	.434653	− 1.833333	2.267986
2.10	.432601	− 2.299099	2.731700
2.20	.430753	− 2.928000	3.358753
2.30	.429067	− 3.825287	4.254354
2.40	.427514	− 5.211111	5.638625
2.50	.426069	− 7.636364	8.062433
2.60	.424714	− 12.977778	13.402491
2.70	.423433	− 34.453333	34.876766
2.80	.422215	71.100000	− 70.677786
2.90	421050	18.884848	− 18.463799
3.00	.419929	11.333333	− 10.913404
3.10	.418848	8.305618	− 7.886770
3.20	.417800	6.673333	− 6.255534
3.30	.416780	5.652288	− 5.235507
3.40	.415786	4.953191	− 4.537405
3.50	.414815	4.444444	− 4.029630
3.60	.413863	4.057576	− 3.643713
3.70	.412929	3.753443	− 3.340514
3.80	.412011	3.508046	− 3.096035

Sigma	$K(\sigma, .0, .00)$	$Y(\sigma, .0, .00)$	$Z(\sigma, .0, .00)$
3.90	.411107	3.305852	− 2.894746
4.00	.410216	3.136364	− 2.726148
4.10	.409336	2.992229	− 2.582893
4.20	.408468	2.868148	− 2.459680
4.30	.407610	2.760202	− 2.352592
4.40	.406762	2.665432	− 2.258670
4.50	.405922	2.581560	− 2.175638
4.60	.405091	2.506806	− 2.101716
4.70	.404267	2.439758	− 2.035490
4.80	.403451	2.379279	− 1.975828
4.90	.402642	2.324449	− 1.921807
5.00	.401840	2.274510	− 1.872670
5.10	.401044	2.228834	− 1.827790
5.20	.400254	2.186897	− 1.786642
5.30	.399471	2.148256	− 1.748786
5.40	.398693	2.112538	− 1.713845
5.50	.397921	2.079422	− 1.681501
5.60	.397154	2.048634	− 1.651479
5.70	.396393	2.019935	− 1.623542
5.80	.395638	1.993120	− 1.597483
5.90	.394887	1.968009	− 1.573122

Sigma	$K(\sigma, .0, 1.00)$	$Y(\sigma, .0, 1.00)$	$Z(\sigma, .0, 1.00)$
1.10	2.901101	82.613299	− 79.712197
1.20	1.651931	26.361277	− 24.709346
1.30	1.235857	14.553953	− 13.318096
1.40	1.027912	9.951707	− 8.923795
1.50	.903123	7.606594	− 6.703471
1.60	.819847	6.216509	− 5.396662
1.70	.760245	5.308432	− 4.548187
1.80	.715408	4.673751	− 3.958343
1.90	.680388	4.207589	− 3.527201
2.00	.652222	3.851977	− 3.199755
2.10	.629027	3.572486	− 2.943459
2.20	.609548	3.347475	− 2.737927
2.30	.592921	3.162710	− 2.569789
2.40	.578526	3.008468	− 2.429942
2.50	.565912	2.877896	− 2.311984
2.60	.554741	2.766034	− 2.211293
2.70	.544754	2.669209	− 2.124455
2.80	.535751	2.584643	− 2.048891
2.90	.527574	2.510201	− 1.982626
3.00	.520097	2.444216	− 1.924119

Sigma	$K(\sigma, .0,1.00)$	$Y(\sigma, .0,1.00)$	$Z(\sigma, .0,1.00)$
3.10	.513217	2.385368	− 1.872151
3.20	.506852	2.332601	− 1.825749
3.30	.500932	2.285056	− 1.784124
3.40	.495399	2.242034	− 1.746634
3.50	.490207	2.202956	− 1.712749
3.60	.485313	2.167341	− 1.682028
3.70	.480683	2.134788	− 1.654106
3.80	.476286	2.104958	− 1.628672
3.90	.472096	2.077564	− 1.605468
4.00	.468091	2.052363	− 1.584272
4.10	.464250	2.029148	− 1.564898
4.20	.460554	2.007742	− 1.547188
4.30	.456988	1.987996	− 1.531009
4.40	.453536	1.969784	− 1.516247
4.50	.450185	1.952997	− 1.502812
4.60	.446922	1.937550	− 1.490628
4.70	.443734	1.923372	− 1.479638
4.80	.440609	1.910411	− 1.469801
4.90	.437535	1.898632	− 1.461097
5.00	.434497	1.888021	− 1.453523
5.10	.431483	1.878587	− 1.447104
5.20	.428476	1.870369	− 1.441893
5.30	.425456	1.863445	− 1.437989
5.40	.422398	1.857954	− 1.435556
5.50	.419269	1.854128	− 1.434859
5.60	.416017	1.852366	− 1.436349
5.70	.412560	1.853401	− 1.440841
5.80	.408734	1.858766	− 1.450032
5.90	.404118	1.872690	− 1.468572

Sigma	$K(\sigma, .0,2.00)$	$Y(\sigma, .0,2.00)$	$Z(\sigma, .0,2.00)$
1.10	3.766836	14.667773	− 10.900937
1.20	2.100052	8.001882	− 5.901830
1.30	1.544147	5.780233	− 4.236087
1.40	1.265827	4.669574	− 3.403747
1.50	1.098460	4.003290	− 2.904829
1.60	.986516	3.559194	− 2.572678
1.70	.906205	3.242072	− 2.335866
1.80	.845641	3.004322	− 2.158681
1.90	.798224	2.819503	− 2.021280
2.00	.759996	2.671751	− 1.911755
2.10	.728442	2.550972	− 1.822530
2.20	.701887	2.450439	− 1.748552

Sigma	$K(\sigma, \quad .0, 2.00)$	$Y(\sigma, \quad .0, 2.00)$	$Z(\sigma, \quad .0, 2.00)$
2.30	.679172	2.365493	− 1.686321
2.40	.659469	2.292810	− 1.633341
2.50	.642173	2.229951	− 1.587778
2.60	.626830	2.175089	− 1.548260
2.70	.613092	2.126827	− 1.513735
2.80	.600691	2.084079	− 1.483388
2.90	.589412	2.045989	− 1.456577
3.00	.579086	2.011874	− 1.432788
3.10	.569575	1.981180	− 1.411605
3.20	.560766	1.953458	− 1.392693
3.30	.552564	1.928336	− 1.375772
3.40	.544892	1.905507	− 1.360614
3.50	.537685	1.884713	− 1.347029
3.60	.530885	1.865741	− 1.334856
3.70	.524444	1.848408	− 1.323964
3.80	.518322	1.832562	− 1.314240
3.90	.512480	1.818074	− 1.305593
4.00	.506889	1.804835	− 1.297947
4.10	.501517	1.792756	− 1.291239
4.20	.496341	1.781762	− 1.285420
4.30	.491337	1.771791	− 1.280454
4.40	.486482	1.762798	− 1.276316
4.50	.481757	1.754748	− 1.272991
4.60	.477141	1.747621	− 1.270480
4.70	.472616	1.741412	− 1.268795
4.80	.468163	1.736128	− 1.267966
4.90	.463760	1.731799	− 1.268039
5.00	.459387	1.728477	− 1.269090
5.10	.455019	1.726242	− 1.271223
5.20	.450628	1.725220	− 1.274593
5.30	.446177	1.725597	− 1.279419
5.40	.441622	1.727653	− 1.286031
5.50	.436897	1.731830	− 1.294933
5.60	.431907	1.738857	− 1.306951
5.70	.426488	1.750053	− 1.323566
5.80	.420322	1.768189	− 1.347867
5.90	.412576	1.801047	− 1.388471

Sigma	$K(\sigma, \quad 2.0, .00)$	$Y(\sigma, \quad 2.0, .00)$	$Z(\sigma, \quad 2.0, .00$
1.10	472805	− .095724	.568529
1.20	.443912	− .177821	.621733
1.30	.415148	− .244681	.659829
1.40	.388282	− .300279	.688560

Sigma	$K(\sigma, \quad 2.0, \quad .00)$	$Y(\sigma, \quad 2.0, \quad .00)$	$Z(\sigma, \quad 2.0, \quad .00)$
1.50	.364414	−.350220	.714634
1.60	.343963	−.399610	.743573
1.70	.326881	−.452675	.779556
1.80	.312878	−.513127	.826005
1.90	.301571	−.584721	.886292
2.00	.292568	−.671879	.964446
2.10	.285507	−.780466	1.065972
2.20	.280073	−.918961	1.199034
2.30	.275997	−1.100513	1.376510
2.40	.273054	−1.346987	1.620041
2.50	271057	−1.697868	1.968925
2.60	.269850	−2.232696	2.502546
2.70	.269305	−3.139297	3.408602
2.80	.269316	−4.993258	5.262574
2.90	.269796	−10.826010	11.095806
3.00	.270674	170.443720	−170.173046
3.10	.271890	10.510868	−10.238978
3.20	.273396	5.677998	−5.404602
3.30	.275153	4.009037	−3.733884
3.40	.277129	3.167739	−2.890609
3.50	.279300	2.663836	−2.384536
3.60	.281643	2.330582	−2.048939
3.70	.284145	2.095712	−1.811567
3.80	.286793	1.922855	−1.636062
3.90	.289578	1.791732	−1.502153
4.00	.292496	1.690169	−1.397674
4.10	.295543	1.610440	−1.314897
4.20	.298719	1.547428	−1.248709
4.30	.302027	1.497636	−1.195609
4.40	.305470	1.458621	−1.153151
4.50	.309055	1.428655	−1.119599
4.60	.312792	1.406517	−1.093725
4.70	.316691	1.391363	−1.074672
4.80	.320765	1.382638	−1.061872
4.90	.325033	1.380031	−1.054999
5.00	.329513	1.383449	−1.053936
5.10	.334228	1.393003	−1.058774
5.20	.339208	1.409025	−1.069817
5.30	.344484	1.432096	−1.087612
5.40	.350095	1.463104	−1.113009
5.50	.356087	1.503333	−1.147246
5.60	.362515	1.554612	−1.192097
5.70	.369446	1.619551	−1.250105
5.80	.376958	1.701927	−1.324969
5.90	.385150	1.807348	−1.422199

Sigma	$K(\sigma, \quad 2.0, 1.00)$	$Y(\sigma, \quad 2.0, 1.00)$	$Z(\sigma, \quad 2.0, 1.00)$
1.10	1.898025	− 4.074725	5.972750
1.20	1.099150	− 2.947422	4.046572
1.30	.816224	− 2.652753	3.468977
1.40	.664932	− 2.599204	3.264136
1.50	.568867	− 2.696442	3.265309
1.60	.502344	− 2.946972	3.449316
1.70	.453980	− 3.407659	3.861639
1.80	.417757	− 4.227583	4.645341
1.90	.390100	− 5.820759	6.210859
2.00	.368711	− 9.789958	10.158669
2.10	.352033	− 33.253184	33.605218
2.20	.338971	23.583045	− 23.244074
2.30	.328727	8.793279	− 8.464552
2.40	.320712	5.472295	− 5.151583
2.50	.314479	4.021776	− 3.707297
2.60	.309685	3.216247	− 2.906562
2.70	.306063	2.708153	− 2.402089
2.80	.303405	2.361261	− 2.057856
2.90	.301545	2.111331	− 1.809786
3.00	.300351	1.924161	− 1.623810
3.10	.299716	1.779904	− 1.480188
3.20	.299556	1.666278	− 1.366722
3.30	.299801	1.575290	− 1.275489
3.40	.300397	1.501534	− 1.201137
3.50	.301298	1.441230	− 1.139932
3.60	.302468	1.391671	− 1.089203
3.70	.303879	1.350876	− 1.046997
3.80	.305508	1.317373	− 1.011865
3.90	.307337	1.290057	− .982720
4.00	.309353	1.268092	− .958739
4.10	.311546	1.250848	− .939302
4.20	.313911	1.237857	− .923946
4.30	.316444	1.228779	− .912335
4.40	.319144	1.223380	− .904235
4.50	.322015	1.221522	− .889507
4.60	325060	1.223155	− .898095
4.70	.328288	1.228311	− .900023
4.80	.331708	1.237108	− .905400
4.90	.335333	1.249760	− .914427
5.00	.339179	1.266588	− .927409
5.10	.343265	1.288045	− .944780
5.20	.347613	1.314746	− .967133
5.30	.352250	1.347523	− .995272
5.40	.357207	1.387492	− 1.030285
5.50	.362519	1.436176	− 1.073657

Sigma	$K(\sigma, \quad 2.0,1.00)$	$Y(\sigma, \quad 2.0,1.00)$	$Z(\sigma, \quad 2.0,1.00)$
5.60	.368223	1.495677	− 1.127454
5.70	.374359	1.568986	− 1.194627
5.80	.380955	1.660544	− 1.279589
5.90	.387974	1.777559	− 1.389585

Sigma	$K(\sigma, \quad 2.0,2.00)$	$Y(\sigma, \quad 2.0,2.00)$	$Z(\sigma, \quad 2.0,2.00)$
1.10	2.696802	− 102.755872	105.452674
1.20	1.484269	4169.630076	− 4168.145807
1.30	1.061340	43.580264	− 42.518925
1.40	.839653	17.822845	− 16.983192
1.50	.701689	10.182325	− 9.480636
1.60	.607805	6.779691	− 6.171886
1.70	.540478	4.951192	− 4.410713
1.80	.490541	3.852232	− 3.361691
1.90	.452631	3.139814	− 2.687182
2.00	.423366	2.651812	− 2.228446
2.10	.400496	2.303183	− 1.902686
2.20	.382465	2.045748	− 1.663283
2.30	.368163	1.850539	− 1.482376
2.40	.356778	1.699280	− 1.342502
2.50	.347705	1.579979	− 1.232274
2.60	.340486	1.484505	− 1.144020
2.70	.334767	1.407186	− 1.072419
2.80	.330275	1.343972	− 1.013697
2.90	.326794	1.291912	− .965118
3.00	.324152	1.248815	− .924663
3.10	.322215	1.213036	− .890820
3.20	.320873	1.183317	− .862444
3.30	.320037	1.158693	− .838655
3.40	.319638	1.138414	− .818777
3.50	.319617	1.121899	− .802282
3.60	.319928	1.108692	− .788764
3.70	.320533	1.098442	− .777908
3.80	.321404	1.090878	− .769474
3.90	.322515	1.085798	− .763283
4.00	.323849	1.083059	− .759210
4.10	.325391	1.082566	− .757175
4.20	.327131	1.084271	− .757140
4.30	.329062	1.088170	− .759107
4.40	.331181	1.094299	− .763118
4.50	.333487	1.102743	− .769255
4.60	.335982	1.113631	− .777648
4.70	.338671	1.127148	− .788477

Sigma	$K(\sigma,\ \ 2.0,2.00)$	$Y(\sigma,\ \ 2.0,2.00)$	$Z(\sigma,\ \ 2.0,2.00)$
4.80	.341560	1.143541	− .801982
4.90	.344659	1.163137	− .818478
5.00	.347981	1.186356	− .838375
5.10	.351541	1.213740	− .862199
5.20	.355357	1.245993	− .890636
5.30	.359450	1.284036	− .924586
5.40	.363842	1.329087	− .965245
5.50	.368558	1.382794	− 1.014236
5.60	.373621	1.447441	− 1.073819
5.70	.379043	1.526309	− 1.147266
5.80	.384799	1.624412	− 1.239613
5.90	.390721	1.750453	− 1.359732

Sigma	$K(\sigma,\ \ 100.0,\ .00)$	$Y(\sigma,\ \ 100.0,\ .00)$	$Z(\sigma,\ \ 100.0,\ .00)$
1.10	.057614	− .006053	.063667
1.20	.067018	− .014593	.081611
1.30	.085539	− .029223	.114763
1.40	.102574	− .049107	.151681
1.50	.117544	− .074272	.191815
1.60	.130718	− .105174	.235892
1.70	.142419	− .142647	.285066
1.80	.152918	− .187951	.340869
1.90	.162428	− .242903	.405331
2.00	.171117	− .310112	.481229
2.10	.179120	− .393376	.572495
2.20	.186541	− .498383	.684924
2.30	.193468	− .633992	.827460
2.40	.199971	− .814745	1.014716
2.50	.206110	− 1.066230	1.272340
2.60	.211934	− 1.437961	1.649895
2.70	.217484	− 2.039778	2.257262
2.80	.222797	− 3.173787	3.396584
2.90	.227905	− 6.081884	6.309789
3.00	.232835	− 29.376025	29.608860
3.10	.237612	12.599850	− 12.362239
3.20	.242257	5.605115	− 5.362857
3.30	.246791	3.771171	− 3.524381
3.40	.251231	2.929722	− 2.678491
3.50	.255596	2.449459	− 2.193863
3.60	.259900	2.141049	− 1.881150
3.70	.264159	1.927999	− 1.663840
3.80	.268389	1.773517	− 1.505128
3.90	.272603	1.657738	− 1.385136

Sigma	$K(\sigma,\ \ 100.0,\ .00)$	$Y(\sigma,\ \ 100.0,\ .00)$	$Z(\sigma,\ \ 100.0,\ .00)$
4.00	.276815	1.569021	− 1.292206
4.10	.281041	1.500115	− 1.219074
4.20	.285295	1.446292	− 1.160997
4.30	.289593	1.404369	− 1.114776
4.40	.293950	1.372148	− 1.078198
4.50	.298383	1.348101	− 1.049718
4.60	.302911	1.331169	− 1.028258
4.70	.307554	1.320639	− 1.013085
4.80	.312333	1.316072	− 1.003739
4.90	.317273	1.317260	− .999986
5.00	.322402	1.324200	− 1.001798
5.10	.327749	1.337099	− 1.009350
5.20	.333350	1.356389	− 1.023038
5.30	.339247	1.382766	− 1.043519
5.40	.345486	1.417261	− 1.071775
5.50	.352122	1.461349	− 1.109227
5.60	.359224	1.517132	− 1.157909
5.70	.366868	1.587626	− 1.220758
5.80	.375153	1.677251	− 1.302098
5.90	.384195	1.792684	− 1.408489

Sigma	$K(\sigma,\ \ 100.0,1.00)$	$Y(\sigma,\ \ 100.0,1.00)$	$Z(\sigma,\ \ 100.0,1.00)$
1.10	.062959	− .007368	.070328
1.20	.069968	− .016209	.086177
1.30	.087941	− .031471	.119412
1.40	.104653	− .052086	.156739
1.50	.119385	− .078086	.197471
1.60	.132371	− .109962	.242333
1.70	.143919	− .148599	.292518
1.80	.154288	− .195330	.349618
1.90	.163687	− .252076	.415764
2.00	.172281	− .321603	.493884
2.10	.180200	− .407954	.588154
2.20	.187547	− .517220	.704767
2.30	.194408	− .658959	.853367
2.40	.200852	− .849020	1.049872
2.50	.206937	− 1.115666	1.322602
2.60	.212711	− 1.514621	1.727332
2.70	.218217	− 2.173023	2.391240
2.80	.223489	− 3.457539	3.681028
2.90	.228559	− 7.040135	7.268694
3.00	.233454	− 68.641660	68.875114
3.10	.238198	10.374816	− 10.136618

Sigma	$K(\sigma,\ 100.0, 1.00)$	$Y(\sigma,\ 100.0, 1.00)$	$Z(\sigma,\ 100.0, 1.00)$
3.20	.242813	5.172091	− 4.929279
3.30	.247318	3.592965	− 3.345647
3.40	.251731	2.833318	− 2.581587
3.50	.256071	2.389295	− 2.133224
3.60	.260351	2.100054	− 1.839703
3.70	.264588	1.898356	− 1.633768
3.80	.268796	1.751146	− 1.482350
3.90	.272990	1.640304	− 1.367314
4.00	.277183	1.555090	− 1.277907
4.10	.281391	1.488759	− 1.207368
4.20	.285627	1.436885	− 1.151257
4.30	.289908	1.396470	− 1.106562
4.40	.294248	1.365441	− 1.071193
4.50	.298666	1.342354	− 1.043688
4.60	.303179	1.326204	− 1.023026
4.70	.307807	1.316322	− 1.008516
4.80	.312571	1.312299	− .999727
4.90	.317497	1.313946	− .996448
5.00	.322611	1.321280	− .998668
5.10	.327945	1.334520	− 1.006575
5.20	.333532	1.354109	− 1.020577
5.30	.339414	1.380751	− 1.041337
5.40	.345639	1.415486	− 1.069847
5.50	.352261	1.459794	− 1.107534
5.60	.359346	1.515786	− 1.156440
5.70	.366973	1.586487	− 1.219514
5.80	.375238	1.676331	− 1.301093
5.90	.384255	1.792033	− 1.407778

Sigma	$K(\sigma,\ 100.0, 2.00)$	$Y(\sigma,\ 100.0, 2.00)$	$Z(\sigma,\ 100.0, 2.00)$
1.10	.068213	− .008818	.077030
1.20	.072873	− .017918	.090791
1.30	.090310	− .033819	.124129
1.40	.106705	− .055181	.161886
1.50	.121205	− .082039	.203243
1.60	.134007	− .114920	.248927
1.70	.145403	− .154764	.300167
1.80	.155645	− .202980	.358625
1.90	.164936	− .261602	.426537
2.00	.173436	− .333563	.506998
2.10	.181271	− .423173	.604445
2.20	.188545	− .536965	.725511
2.30	.195341	− .685271	.880612

Appendix C

Sigma	$K(\sigma, \quad 100.0, 2.00)$	$Y(\sigma, \quad 100.0, 2.00)$	$Z(\sigma, \quad 100.0, 2.00)$
2.40	.201726	− .885404	1.087130
2.50	.207758	− 1.168683	1.376440
2.60	.213484	− 1.598102	1.811586
2.70	.218945	− 2.321831	2.540776
2.80	.224177	− 3.790419	4.014596
2.90	.229209	− 8.330862	8.560071
3.00	.234069	217.448299	− 217.214230
3.10	.238781	8.837337	− 8.598556
3.20	.243365	4.806003	− 4.562637
3.30	.247842	3.432892	− 3.185050
3.40	.252230	2.744153	− 2.491923
3.50	.256544	2.332680	− 2.076137
3.60	.260801	2.061036	− 1.800235
3.70	.265015	1.869913	− 1.604898
3.80	.269202	1.729551	− 1.460348
3.90	.273376	1.623395	− 1.350019
4.00	.277550	1.541528	− 1.263978
4.10	.281739	1.477670	− 1.195930
4.20	.285958	1.427674	− 1.141715
4.30	.290222	1.388719	− 1.098497
4.40	.294546	1.358849	− 1.064303
4.50	.298948	1.336695	− 1.037747
4.60	.303446	1.321309	− 1.017863
4.70	.308059	1.312060	− 1.004002
4.80	.312809	1.308568	− .995759
4.90	.317721	1.310666	− .992946
5.00	.322821	1.318387	− .995566
5.10	.328140	1.331963	− 1.003823
5.20	.333714	1.351846	− 1.018133
5.30	.339582	1.378751	− 1.039169
5.40	.345792	1.413721	− 1.067929
5.50	.352399	1.458248	− 1.105849
5.60	.359468	1.514446	− 1.154978
5.70	.367078	1.585352	− 1.218274
5.80	.375323	1.675415	− 1.300092
5.90	.384315	1.791384	− 1.407069

Appendix D

The tables refer to the evaluation of

$$F_0(\sigma, m_1) = \int_{1.01}^{\sigma} \left[\frac{(\theta^2 - 1) + m_1^2 \gamma [3\theta - 1 - 6\theta x + 3(1 + \theta) x^2 - 2 x^3]/2}{(\theta - x)(x - 1)[\theta + 1 + \gamma m_1^2 (x - 1)^2/2] K_2(x, m_1)} \right] dx.$$

The numbers in parentheses refer to the power of ten by which each entry is to be multiplied.

σ	$F_0(\sigma, 0)$	$F_0(\sigma, 1)$	σ	$F_0(\sigma, 2)$	$F_0(\sigma, 3)$
1.2	6.65000 (0)	6.66275 (0)	1.2	6.70180 (0)	6.76868 (0)
1.4	1.36500 (1)	1.37038 (1)	1.4	1.38736 (1)	1.41736 (1)
1.6	2.06500 (1)	2.07765 (1)	1.6	2.11842 (1)	2.18986 (1)
1.8	2.76500 (1)	2.78834 (1)	1.8	2.86366 (1)	2.98974 (1)
2.0	3.46500 (1)	3.50263 (1)	2.0	3.62160 (1)	3.80758 (1)
2.2	4.16500 (1)	4.22054 (1)	2.2	4.38964 (1)	4.63410 (1)
2.4	4.86500 (1)	4.94203 (1)	2.4	5.16488 (1)	5.46244 (1)
2.6	5.56500 (1)	5.66694 (1)	2.6	5.94468 (1)	6.28849 (1)
2.8	6.26500 (1)	6.39504 (1)	2.8	6.72696 (1)	7.11011 (1)
3.0	6.96500 (1)	7.12610 (1)	3.0	7.51020 (1)	7.92647 (1)
3.2	7.66500 (1)	7.85988 (1)	3.2	8.29343 (1)	8.73751 (1)
3.4	8.36500 (1)	8.59614 (1)	3.4	9.07609 (1)	9.54358 (1)
3.6	9.06500 (1)	9.33473 (1)	3.6	9.85792 (1)	1.03452 (2)
3.8	9.76500 (1)	1.00755 (2)	3.8	1.06390 (2)	1.11432 (2)
4.0	1.04650 (2)	1.08185 (2)	4.0	1.14194 (2)	1.19383 (2)
4.2	1.11650 (2)	1.15638 (2)	4.2	1.22000 (2)	1.27314 (2)
4.4	1.18650 (2)	1.23116 (2)	4.4	1.29803 (2)	1.35234 (2)
4.6	1.25650 (2)	1.30623 (2)	4.6	1.37623 (2)	1.43155 (2)
4.8	1.32650 (2)	1.38167 (2)	4.8	1.45467 (2)	1.51093 (2)
5.0	1.39650 (2)	1.45761 (2)	5.0	1.53355 (2)	1.59070 (2)
5.2	1.46500 (2)	1.53426 (2)	5.2	1.61318 (2)	1.67117 (2)
5.4	1.53650 (2)	1.61206 (2)	5.4	1.69410 (2)	1.75300 (2)
5.6	1.60650 (2)	1.69194 (2)	5.6	1.77755 (2)	1.83738 (2)
5.8	1.67650 (2)	1.77700 (2)	5.8	1.86748 (2)	1.92857 (2)

Appendix E

Tables are presented for the parameters

$$K_c(\sigma, m_1),$$
$$K_s(\sigma, m_1)$$

and the functions

$$F_c(\sigma, m_1),$$
$$F_s(\sigma, m_1).$$

Sigma	$K_c(\sigma, \ .0)$	$K_s(\sigma, \ .0)$	$F_c(\sigma, \ .0)$	$F_s(\sigma, \ .0)$
1.10	.494525	.483770	4.813653	4.860431
1.20	.488343	.473108	6.266187	6.352881
1.30	.481723	.465952	7.145358	7.261529
1.40	.474846	.461024	7.791676	7.928761
1.50	.467836	.457516	8.311639	8.462564
1.60	.460780	.454905	8.752652	8.911513
1.70	.453738	.452849	9.140005	9.301820
1.80	.446754	.451122	9.488811	9.649325
1.90	.439858	.449577	9.808841	9.964376
2.00	.433071	.448113	10.106796	10.254127
2.10	.426408	.446667	10.387487	10.523751
2.20	.419879	.445197	10.654499	10.777120
2.30	.413490	.443677	10.910591	11.017219
2.40	.407245	.442092	11.157946	11.246410
2.50	.401146	.440433	11.398334	11.466600
2.60	.395193	.438699	11.633226	11.679365
2.70	.389384	.436888	11.863870	11.886029
2.80	.383719	.435005	12.091348	12.087724
2.90	.378194	.433052	12.316618	12.285441
3.00	.372807	.431034	12.540547	12.480056
3.10	.367555	.428958	12.763936	12.672361
3.20	.362434	.426828	12.987540	12.863088
3.30	.357442	.424649	13.212086	13.052920
3.40	.352573	.422427	13.438287	13.242512
3.50	.347826	.420168	13.666860	13.432503
3.60	.343196	.417876	13.898536	13.623525
3.70	.338680	.415555	14.134075	13.816223
3.80	.334274	.413210	14.374280	14.011260

Sigma	$K_c (\sigma, .0)$	$K_s (\sigma, .0)$	$F_c (\sigma, .0)$	$F_s (\sigma, .0)$
3.90	.329975	.410846	14.620019	14.209334
4.00	.325779	.408465	14.872238	14.411193
4.10	.321684	.406071	15.131985	14.617652
4.20	.317685	.403667	15.400442	14.829615
4.30	.313781	.401257	15.678956	15.048100
4.40	.309968	.398843	15.969086	15.274271
4.50	.306243	.396427	16.272662	15.509483
4.60	.302603	.394011	16.591862	15.755339
4.70	.299046	.391598	16.929321	16.013764
4.80	.295569	.389190	17.288280	16.287117
4.90	.292169	.386788	17.672801	16.578342
5.00	.288845	.384393	18.088080	16.891199
5.10	.285593	.382007	18.540938	17.230607
5.20	.282412	.379632	19.040583	17.603201
5.30	.279299	.377268	19.599898	18.018252
5.40	.276253	.374917	20.237726	18.489306
5.50	.273271	.372578	20.983248	19.037332
5.60	.270351	.370254	21.885321	19.697424
5.70	.267492	.367945	23.035515	20.535341
5.80	.264691	.365651	24.639446	21.698719
5.90	.261948	.363373	27.354146	23.659287
5.99	.259526	.361337	36.564343	30.281899
6.00	.259259	.361111		

Sigma	$K_c (\sigma, 1.0)$	$K_s (\sigma, 1.0)$	$F_c (\sigma, 1.0)$	$F_s (\sigma, 1.0)$
1.10	.491342	.479356	4.825388	4.878160
1.20	.480093	.463116	6.300004	6.398577
1.30	.467418	.449501	7.210155	7.343951
1.40	.454160	.437543	7.895836	8.056558
1.50	.440910	.426741	8.463102	8.644147
1.60	.428064	.416848	8.958951	9.154972
1.70	.415868	.407752	9.408287	9.614863
1.80	.404465	.399402	9.825842	10.039239
1.90	.393915	.391771	10.221007	10.437998
2.00	.384231	.384840	10.600098	10.817829
2.10	.375389	.378583	10.967539	11.183429
2.20	.367344	.372973	11.326526	11.538191
2.30	.360040	.367973	11.679427	11.884622
2.40	.353415	.363547	12.028035	12.224613
2.50	.347408	.359652	12.373737	12.559611
2.60	.341956	.356249	12.717625	12.890744
2.70	.337001	.353297	13.060580	13.218908
2.80	.332488	.350756	13.403328	13.544824

Sigma	$K_c\,(\sigma,\ 1.0)$	$K_s\,(\sigma,\ 1.0)$	$F_c\,(\sigma,\ 1.0)$	$F_s\,(\sigma,\ 1.0)$
2.90	.328367	.348589	13.746486	13.869092
3.00	.324593	.346761	14.090597	14.192221
3.10	.321124	.345240	14.436153	14.514662
3.20	.317921	.343994	14.783619	14.836824
3.30	.314952	.342997	15.133450	15.159095
3.40	.312185	.342224	15.486108	15.481857
3.50	.309593	.341653	15.842075	15.805499
3.60	.307153	.341262	16.201867	16.130429
3.70	.304841	.341033	16.566047	16.457084
3.80	.302639	.340951	16.935239	16.785943
3.90	.300528	.341001	17.310148	17.117537
4.00	.298494	.341169	17.691570	17.452468
4.10	.296522	.341445	18.080424	17.791421
4.20	.294600	.341819	18.477771	18.135182
4.30	.292716	.342283	18.884853	18.484669
4.40	.290860	.342829	19.303136	18.840959
4.50	.289023	.343451	19.734366	19.205331
4.60	.287197	.344145	20.180650	19.579322
4.70	.285374	.344906	20.644558	19.964804
4.80	.283546	.345733	21.129277	20.364089
4.90	.281709	.346623	21.638821	20.780084
5.00	.279855	.347576	22.178348	21.216517
5.10	.277980	.348591	22.754643	21.678284
5.20	.276077	.349670	23.376887	22.172002
5.30	.274142	.350815	24.057940	22.706927
5.40	.272169	.352029	24.816625	23.296598
5.50	.270154	.353317	25.682109	23.961979
5.60	.268091	.354685	26.703240	24.738163
5.70	.265976	.356138	27.971578	25.690910
5.80	.263803	.357687	29.692736	26.967860
5.90	.261566	.359340	32.523743	29.040751
5.99	.259493	.360928	41.837829	35.763392
6.00	.259259	.361111		

Sigma	$K_c\,(\sigma,\ 2.0)$	$K_s\,(\sigma,\ 2.0)$	$F_c\,(\sigma,\ 2.0)$	$F_s\,(\sigma,\ 2.0)$
1.10	.486218	.473433	4.844391	4.901225
1.20	.464842	.446201	6.366905	6.476023
1.30	.440781	.419714	7.350684	7.505192
1.40	.417226	.395389	8.133272	8.329138
1.50	.395853	.374088	8.818170	9.052954
1.60	.377297	.356091	9.449649	9.721676
1.70	.361613	.341287	10.049730	10.357560
1.80	.348571	.329359	10.630190	10.972291

Sigma	K_c (σ, 2.0)	K_s (σ, 2.0)	F_c (σ, 2.0)	F_s (σ, 2.0)
1.90	.337841	.319925	11.197483	11.572064
2.00	.329076	.312603	11.755110	12.160043
2.10	.321954	.307046	12.304889	12.737692
2.20	.316190	.302952	12.847676	13.305548
2.30	.311538	.300064	13.383811	13.863674
2.40	.307791	.298166	13.913380	14.411939
2.50	.304777	.297079	14.436379	14.950175
2.60	.302348	296652	14.952810	15.478271
2.70	.300385	.296761	15.462731	15.996214
2.80	.298788	.297303	15.966280	16.504108
2.90	.297474	.298192	16.463690	17.002171
3.00	.296375	.299359	16.955282	17.490730
3.10	.295435	.300744	17.441466	17.970208
3.20	.294609	.302301	17.922732	18.441110
3.30	.293860	.303990	18.399642	18.904013
3.40	.293157	.305780	18.872831	19.359556
3.50	.292476	.307644	19.342997	19.808431
3.60	.291797	.309562	19.810904	20.251382
3.70	.291104	.311517	20.277387	20.689199
3.80	.290386	.313497	20.743352	21.122723
3.90	.289631	.315491	21.209787	21.552850
4.00	.288832	.317492	21.677777	21.980539
4.10	.287983	.319495	22.148521	22.406821
4.20	.287080	.321497	22.623354	22.832821
4.30	.286118	.323495	23.103782	23.259777
4.40	.285095	.325489	23.591523	23.689073
4.50	.284009	.327481	24.088563	24.122277
4.60	.282859	.329473	24.597239	24.561200
4.70	.281642	.331467	25.120335	25.007970
4.80	.280358	.333468	25.661242	25.465140
4.90	.279007	.335481	26.224164	25.935841
5.00	.277586	.337513	26.814438	26.424012
5.10	.276096	.339570	27.439016	26.934745
5.20	.274535	.341660	28.107235	27.474839
5.30	.272901	.343795	28.832101	28.053720
5.40	.271194	.345983	29.632573	28.685084
5.50	.269411	.348239	30.537945	29.390041
5.60	.267550	.350576	31.597182	30.203823
5.70	.265609	.353011	32.901955	31.192314
5.80	.263582	.355563	34.657979	32.503273
5.90	.261468	.358255	37.522379	34.608546
5.99	.259484	.360817	46.865332	41.359025
6.00	.259259	.361111		

Sigma	$K_e\,(\sigma,\ 3.0)$	$K_i\,(\sigma,\ 3.0)$	$F_e\,(\sigma,\ 3.0)$	$F_i\,(\sigma,\ 3.0)$
1.10	.479307	.466114	4.870725	4.929794
1.20	.444566	.424520	6.466919	6.585572
1.30	.408259	.384351	7.565856	7.744583
1.40	.376511	.350503	8.498753	8.741238
1.50	.350954	.324052	9.361724	9.672245
1.60	.331222	.304265	10.190973	10.572853
1.70	.316361	.289919	11.000428	11.455261
1.80	.305383	.279844	11.794412	12.321836
1.90	.297432	.273066	12.573075	13.170893
2.00	.291808	.268816	13.335018	13.999515
2.10	.287956	.266496	14.078589	14.804904
2.20	.285438	.265644	14.802470	15.584963
2.30	.283910	.265902	15.505889	16.338458
2.40	.283104	.266994	16.188637	17.064964
2.50	.282815	.268706	16.850989	17.764724
2.60	.282883	.270871	17.493609	18.438494
2.70	.283186	.273363	18.117440	19.087388
2.80	.283630	.276083	18.723617	19.712760
2.90	.284144	.278954	19.313401	20.316108
3.00	.284675	.281918	19.888123	20.899001
3.10	.285185	.284931	20.449149	21.463033
3.20	.285643	.287959	20.997854	22.009786
3.30	.286028	.290976	21.535612	22.540818
3.40	.286326	.293965	22.063784	23.057644
3.50	.286526	.296911	22.583723	23.561740
3.60	.286622	.299807	23.096773	24.054542
3.70	.286610	.302645	23.604282	24.537451
3.80	.286489	.305424	24.107611	25.011848
3.90	.286258	.308141	24.608149	25.479100
4.00	.285918	.310799	25.107333	25.940580
4.10	.285471	.313398	25.606674	26.397686
4.20	.284919	.315942	26.107779	26.851862
4.30	.284264	.318436	26.612397	27.304628
4.40	.283510	.320883	27.122458	27.757615
4.50	.282657	.323291	27.640139	28.212610
4.60	.281711	.325666	28.167939	28.671616
4.70	.280671	.328015	28.708793	29.136931
4.80	.279542	.330345	29.266221	29.611258
4.90	.278325	.332666	29.844544	30.097863
5.00	.277021	.334987	30.449202	30.600802
5.10	.275632	.337318	31.087238	31.125273
5.20	.274159	.339671	31.768074	31.678167
5.30	.272601	.342057	32.504787	32.268996
5.40	.270960	.344491	33.316403	32.911531
5.50	.269233	.346987	34.232275	33.626948

Sigma	K_c (σ, 3.0)	K_s (σ, 3.0)	F_c (σ, 3.0)	F_s (σ, 3.0)
5.60	.267421	.349564	35.301422	34.450539
5.70	.265520	.352240	36.615563	35.448243
5.80	.263529	.355039	38.380455	36.767868
5.90	.261444	.357986	41.253262	38.881303
5.99	.259482	.360790	50.603417	45.638730
6.00	.259259	.361111		

MATHEMATICS